# THE FOCAL EASY GUIDE TO
# FINAL CUT EXPRESS

## The Focal Easy Guide Series

Focal Easy Guides are the best choice to get you started with new software, whatever your level. Refreshingly simple, they do *not* attempt to cover everything, focusing solely on the essentials needed to get immediate results.

Ideal if you need to learn a new software package quickly, the Focal Easy Guides offer an effective, time-saving introduction to the key tools, not hundreds of pages of confusing reference material. The emphasis is on quickly getting to grips with the software in a practical and accessible way to achieve professional results.

Highly illustrated in color, explanations are short and to the point. Written by professionals in a user-friendly style, the guides assume some computer knowledge and an understanding of the general concepts in the area covered, ensuring they aren't patronizing!

### Series editor: Rick Young (www.digitalproduction.net)

Director and Founding Member of the UK Final Cut User Group, Apple Solutions Expert and freelance television director/editor, Rick has worked for the BBC, Sky, ITN, CNBC and Reuters. Also a Final Cut Pro Consultant and author of the best-selling *The Easy Guide to Final Cut Pro*.

### Titles in the series:

***The Easy Guide to Final Cut Pro 3,*** **Rick Young**

***The Focal Easy Guide to Final Cut Pro 4,*** **Rick Young**

***The Focal Easy Guide to Final Cut Express,*** **Rick Young**

# THE FOCAL EASY GUIDE TO
# FINAL CUT EXPRESS

## For new users and professionals

### RICK YOUNG

AMSTERDAM • BOSTON • HEIDELBERG • LONDON • NEW YORK • OXFORD
PARIS • SAN DIEGO • SAN FRANCISCO • SINGAPORE • SYDNEY • TOKYO

Focal Press is an imprint of Elsevier

Focal Press
An imprint of Elsevier
Linacre House, Jordan Hill, Oxford OX2 8DP
200 Wheeler Road, Burlington MA 01803

First published 2004
Copyright © 2004, Rick Young. All rights reserved

The right of Rick Young to be identified as the author of this work
has been asserted in accordance with the Copyright, Designs and
Patents Act 1988

No part of this publication may be reproduced in any material form (including
photocopying or storing in any medium by electronic means and whether
or not transiently or incidentally to some other use of this publication) without
the written permission of the copyright holder except in accordance with the
provisions of the Copyright, Designs and Patents Act 1988 or under the terms of
a licence issued by the Copyright Licensing Agency Ltd, 90 Tottenham Court Road,
London, England W1T 4LP. Applications for the copyright holder's written
permission to reproduce any part of this publication should be addressed
to the publisher

Permissions may be sought directly from Elsevier's Science and Technology Rights
Department in Oxford, UK: phone: (+44) (0) 1865 843830; fax: (+44) (0) 1865 853333;
e-mail: permissions@elsevier.co.uk. You may also complete your request on-line via the
Elsevier homepage (www.elsevier.com), by selecting 'Customer Support'
and then 'Obtaining Permissions'

**British Library Cataloguing in Publication Data**
A catalogue record for this book is available from the British Library

**Library of Congress Cataloguing in Publication Data**
A catalogue record for this book is available from the Library of Congress

ISBN   0 240 51946 9

For information on all Focal Press publications visit our website at
www.focalpress.com

Typeset by Newgen Imaging Systems (P) Ltd, Chennai, India
Printed and bound in Italy

# Contents

**Preface**   ix
**For New Users**   xi

## The Old and the New   1
   The Digital Laboratory   3
   Inside Your Mac   4
   Hardware and Software Requirements   5
   How Much Hard Drive Space?   6
   Decks and Cameras   6
   Firewire   7
   A World of Different Standards   8
   Television Aspect Ratio   9

## Hands On   11
   Loading the Software   12
   Initial Setup   12

## The Interface   15
   Arranging the Interface   17
   Learning a Custom Layout   20
   Building a Film Structure   20
   Important Details About the Interface   21
   Saving Projects   23

## Setup and Capture   25
   DV Audio   26
   Easy Setup   27
   Setting Scratch Disks   28
   Working with Formats other than DV   30

Methods of Capturing DV Footage   31

Deck Control   31

The Capture Window   32

Capture Clip   34

Capture Now   35

Working Around the Lack of Batch Capture   36

Getting the Most Out of the Capture Process   37

Dealing with Timecode Breaks   37

Importing Music from CD   38

Converting Audio Sample Rates   40

## Sorting Through Your Footage   43

Viewing Clips   44

Playing Video Through Firewire   44

Setting Poster Frames   45

DV Start/Stop Detection   46

Working with Bins   48

Working in List Mode   51

Searching for Clips   52

## The Cutting Room   55

Insert and Overwrite Editing   56

Getting Started with Editing   57

Distinguishing Between Insert and Overwrite   60

Three Point Editing   64

Other Editing Options   65

Modifying 'In' and 'Out' Points   66

Directing the Flow of Audio/Video   67

Locking Tracks   68

Adding and Deleting Tracks   69

Essential Editing Tools   69

Undo/Redo   71

Linked/Unlinked Selection   71

Moving Edits in the Timeline   73

Selecting Multiple Items in the Timeline   75

Edit Like Using a Word Processor   77

Snapping and Skipping Between Shots   77

The Razorblade Tool   78

Magnifier Tool   79

Bringing Clips Back into Sync   80

Creating New Sequences   81

Slow/Fast Motion   83

Rendering   84

Using the Real-Time Effects   86

Subclips   87

Freeze Frame   88

Match Frame Editing   89

Split Edits   91

Drag and Drop Editing   94

Extending/Reducing Clips by Dragging   97

# Effects   99

Single and Multi-Layered Effects   100

The Concept of Media Limit (Handles)   102

Applying Transitions   103

Changing Transition Durations   103

Applying Filters   104

Compositing   107

Methods of Creating Multiple Tracks   108

The Motion Tab   109

Using the Motion Tab   109

Image and Wireframe   112

# CONTENTS

Titlesafe   113

Working with Multi-Layers   114

Keyframing Images   117

Multi-Layered Dissolves   120

Copy and Pasting Attributes   122

Titling   123

## Working with Audio   129

Setting Correct Audio Levels   130

Getting the Most Out of Your Audio   130

Converting Clips into Stereo Pairs   131

Adjusting Audio Levels   132

Adding Sound Fades   133

Adding Audio Cross Fades   135

Adding Audio Tracks   136

Mixdown Audio   138

## Output   139

Playing it Safe   140

Print to Video   140

Other Forms of Distribution   142

## Final Cut Pro 4   147

Final Cut Pro vs Final Cut Express   148

**Epilog**   151
**Index**   153

# Preface

Final Cut Express is an extraordinarily good editor.

It is extremely reliable and stable.

It has a great interface.

It does what it says it will do.

Final Cut Express is way ahead of iMovie. The two don't compare. If you're coming from an iMovie background get ready to drive a real editing suite.

Suite Drivers they used to call us back in the '80s and '90s, before non-linear took over. We were the ones who operated the on-line edit suites of television stations and post-houses – we were highly skilled, highly paid.

Three, Four, Five, Six machine edit suites packed with DVEs, character generators, audio mixers, vision mixers, racks, patch bays, graphic suites just down the corridor.

That's what you are dealing with in Final Cut Express. Everything is now inside a single box. An editing system capable of delivering fully professional results, reliable and stable, at your command.

# PREFACE

The aim of this book is teach the essential knowledge one needs to operate Final Cut Express. For those new to editing, learning any computer-based system is going to be challenging; for those who are experienced with other editing systems, learning a new system requires dedication and effort. This book aims to teach two things:

**(i)** an overview of the post-production process

**(ii)** the technical skills needed to use Final Cut Express.

Final Cut Express is closely related to the full version of Final Cut Pro; however, it is aimed at the DV film-maker rather than users of high-end professional formats. This does not mean Final Cut Express cannot be used for professional work. Final Cut Express provides all the tools and features one needs to work at a professional or prosumer level (providing the source material is acquired or transferred to DV tape).

Don't be fooled by the inexpensive price which Final Cut Express sells for. I repeat – Final Cut Express is well capable of producing professional results, at the DV level. The challenge for the editor is to use this software creatively and to work within its limitations. Many so-called professional software packages, only a few years ago, would have sold for many times the cost of a single copy of Final Cut Express and provided substantially fewer features, worse quality and a less professional interface and workflow.

So read on and learn – if you get bored with all the film theory, skip the first few chapters and dive straight into Setup and Capture. This is the launch point for a condensed power-house of how to edit with Final Cut Express from both a practical and technical perspective. This is where the real action starts.

**Rick Young**
*Producer/Director/Editor*
*London, UK*

# For New Users

Welcome to the world of editing. You are in for a real treat. Final Cut Express is without doubt a great editor to work with. It allows the person driving the interface to work with precision and power. You can do everything from cutting, to sound mixing, titles, effects, compositing, ... the list goes on. Eighty years ago Eisenstein (more of him later) was cutting with a blade – we now cut with tools which are infinitely more effective.

Do not be afraid of this technology – it works and it is not hard to use. You will need to master a few techniques and you will then be well on your way to producing films which are truly professional in structure, technique and their ability to tell stories.

This book will teach you the essential techniques you need to know to edit a film using Final Cut Express. It will not teach you how to edit. This is a skill which needs to be developed by you as you come to know the program.

It's been said before, 'you either have it or you don't'.

It's also been said, 'anyone can edit, you could train a monkey to edit'.

I've also heard the phrase 'everything cuts!'

So when it comes to the editing process rest assured, with perseverence you will be able to knock out films using Final Cut Express.

Sergei Eisenstein only made six films in his lifetime.

You will be able to make as many films as you are prepared to put the time into.

But first you need to know the basic principles of how non-linear editing works.

# THE OLD AND THE NEW

It all started with a blade, in dimly lit rooms with chinagraph pencils and rickety hand-driven winders. Film editing was originally a mechanical process which was achieved by physically cutting and splicing the film. A type of cement glue was used to join the pieces of celluloid together; later the glue was replaced with material similar to sticky tape. There were no dissolves, no effects, just cuts. And masterpieces were created.

In 1925 Russian film-maker Sergei Eisenstein created the *Battleship Potempkin*, a film which has been described as one of the most influential films ever made. In a single film Eisenstein showed how time could be collapsed and expanded purely through editing technique. Barely anyone who has attended film school will have escaped studying Eisenstein's work. Yet this pioneer film-maker, who has been referred to as the father of montage, used the simplest of tools to create his masterpiece.

It may have all started with a blade, however, today editing takes place on sophisticated work stations with more power than the computers used to send men to the moon! And the cost of these work stations has plummeted to an all time low. With nothing more than a DV camera, Firewire Mac and editing software, such as Final Cut Express, one has tools which are infinitely superior to film-makers of previous generations.

No matter how great the tools, it is essential to understand the principles of film-making.

So before we go any further let's step back in time and understand the film-making process the way it used to be… .

## The Digital Laboratory

Think of your computer loaded with Final Cut Express as being like a digital laboratory. In the days when cine-film was the only means for movie-making, everyone relied on the lab. Film would be processed at the lab; effects created; titles added; there were work prints; answer prints; release prints... the lab was central to virtually every facet of the post-production process.

Your Mac is a digital lab just waiting for you to stir the potions.

When working with any kind of videotape it is no longer necessary to develop the film. However, it is necessary to go through a set of processes which can be likened to the old way of working.

While film needed to be developed the images recorded on videotape need to be transferred from tape to hard drive – this process is known as **capture**.

The raw material must then be ordered and structured. In the film world this would take place in the cutting room; when using Final Cut Express an electronic equivalent to the cutting room is provided in the layout of the interface. It is here that the **editing** takes place.

Once the picture was edited the sound must be **mixed**. Dubbing suites with many machines running in synchronization were traditionally used. Inside your computer multiple tracks are stacked visually to represent what would have once required a room full of equipment.

Effects and titles were traditionally created using a device known as an optical printer. Film exposed in the optical printer would then be immersed in developing tanks, in total darkness, to emerge, as if by magic, with hundreds of tiny transparent images; when projected these would light up a room... Final Cut Express uses electronic processes to achieve these results. Video tracks are layered in order of priority to build effects which can be made up of many different layers. This process is known as **compositing**.

Finally, the original negative would be cut and matched by technicians, wearing pure white gloves, in dust-proof rooms. Release prints were produced so the film could be distributed to cinemas and later television stations throughout the world. Release prints in the modern world are recorded onto digital tape, DVD or the final edit may need to be prepared for CD-ROM or internet delivery. This phase of the process is known as **output**.

It should be obvious that a distinct set of processes takes place in the editing of any production. When using Final Cut Express these processes can be broken down into five areas.

(i) capture (ii) editing (iii) sound mixing (iv) compositing (v) output

**Capture**

**Editing**

**Sound Mixing**

**Compositing**

**Output**

Learn how to perform these essential tasks and you will be well armed with the knowledge needed to edit any program. Once these processes are learnt, you, as the editor, will be able to concentrate on the creative aspects of the editing process. Only when one moves beyond the mechanics of the editing can Final Cut Express be used to its full potential.

## Inside Your Mac

Your Mac is made up of a bunch of components, specifically engineered to work together. There are hard drives, fans, a motherboard, memory, circuits, a power supply, ports and slots. Data pumps through the internal system while the keyboard and mouse act as the interface between the computer and the mind of the operator. While it is not essential to understand exactly what goes on inside your Mac it is helpful to have a general overview – particularly with regards to memory and available hard drive space. These two areas are critical to having an efficient and well-managed machine.

## Hardware and Software Requirements

The hardware you need to use Final Cut Express:

**1** A Firewire camera or DV Deck

**2** A G3, G4 or G5 Firewire Mac with at least 256 MB of ram and enough free hard drive space to store your video files. To access the Real-Time effects you will need a G4 or G5 processor running at 500 MHz or above with at least 384 MB of ram.

You must use OSX and I would strongly recommend OS 10.2 or above.

Your Digital Video Deck or Camera will need DV 'in' and 'out'. Ideally you should also have a television monitor and a pair of external speakers.

## How Much Hard Drive Space?

Basically, the more the better. Obviously every one of us is on some sort of limit and you've got to stop somewhere. With that in mind, I'd recommend at least a 60 gigabyte drive. This will add up to, even with software loaded, at least 50 gigabytes of free space – which will give you about 4 hours of storage.

Video at DV resolution chews up approximately one gigabyte to four and a half minutes of sound and video. It is easy, therefore, to work out how much material you can store on hard disk. Simply multiply the capacity of your hard drive by 4.5 and then divide the result by 60. This will calculate the amount of storage you will get in hours and minutes. The measurement of 4.5 minutes to the gigabyte is a conservative estimate. You actually get slightly more.

A lot has been written over the years of the benefits of working with the operating system of your computer on one drive and storing your captured clips on a separate drive. This is really the best way to configure your system, but in the real world, a lot of people will have to use a single drive for the operating system and media storage for the simple reason that they only have one hard drive physically installed inside their computer.

Just be aware if you do use the same drive for both the operating system and media files, the risk of dropped frames will increase. This may or may not present a problem.

## Decks and Cameras

The digital video revolution all started with one camera: the Sony VX-1000. When this camera appeared on the market in 1996 the world went crazy. I remember the BBC had purchased 100 of these and the camera had only just been released. Then I started hearing that the BBC had a VX-1000 in every single department in the whole of the BBC. Documentaries were filmed with this camera, multi-camera shoots were produced and the professional world with all their big cameras sat back in astonishment as the world of acquisition was redefined, apparently, overnight!

Shortly after the VX-1000 was released another, less-famous, Sony product appeared on the scene. This was the DHR-1000 – a Digital Video Editing deck with a bi-directional Firewire port.

This was just the sort of technology the world wanted desperately. Finally, a low cost, lossless, high-quality camera/editing solution had arrived. This exact same technology forms the basis of Firewire editing systems today – only the deck or camera is usually connected to a computer rather than editing from camera to deck or vice-versa.

When using a Firewire-based editing system you can work with either a camera or deck – providing the camera has both Firewire 'in' and 'out'. The advantages of having a deck are: (i) you don't beat up your camera every time you capture footage (ii) a deck offers other features such as different inputs, the ability to work with large or small size tapes, a large timecode display, a jog/shuttle wheel, and often a built-in edit controller.

## Firewire

The golden rule when connecting your Mac and camera/deck with a Firewire cable is to make sure the connector is the correct size. Without sounding too basic or elementary, make sure you insert the Firewire connector correctly – if you jam it in backwards you will end up with a bent Firewire port.

Firewire ports are identified by a symbol (which looks remarkably similar to a nuclear warning symbol).

Firewire cables come in several forms: either a 6-pin (large) or 4-pin (small) connector may be used. Cables can be made up of any combination of small to small, large to large, or small to large connectors. The larger 6-pin Firewire port is found on the back or side of your Mac (depending on which Mac you have), while the smaller 4-pin Firewire connector is located on your camera or deck. Recently, a new version of Firewire, known as Firewire 2, has been released. Firewire 2 has a maximum transfer speed of 800 MB per second which is twice the speed of the original version, known as Firewire 400.

4-Pin/6-Pin Firewire Connectors

6-Pin Firewire Port

4-Pin Firewire Port

Simply plug the large end into your Mac and the small end into your camera or deck. Firewire cables are hot-pluggable which means they can be connected or disconnected while the Mac is switched on or off, although, ideally, the devices should be plugged together prior to launching Final Cut Express. Otherwise a warning message will appear to alert you to the fact that no Firewire device is being seen.

## A World of Different Standards

The world we live in is full of conflicting standards: some countries use 120 volts for power, others 240 volts; vehicles drive on the right-hand side of the road in some countries, in others on the left; there are numerous currencies for financial transactions from country to country. When it comes to television production different standards exist.

DV-NTSC applies to the USA, Japan and many other parts of the world; whereas DV-PAL is used through most of Europe, Australia, and parts of Asia. There are other formats such as SECAM which is used in France and North Africa, and variations of PAL and NTSC are used in South America. However, PAL and NTSC remain the dominant formats.

It is simple but crucial to set the correct video format when working with Final Cut Express.

## Television Aspect Ratio

Another consideration is whether the footage you are working with has been filmed in widescreen anamorphic (16 × 9) or standard television format (4 × 3).

**Widescreen Anamorphic**

**Standard 4 × 3 Television Format**

**Cropped 4 × 3 or Letterbox**

Do not confuse letterbox (cropped 4 × 3) with true widescreen. Most consumer cameras do not offer a true widescreen anamorphic mode of operation. However, many offer a cropped 4 × 3 letterbox setting.

You will need to consult your camera manual to determine the settings available for the camera you are using.

# HANDS ON

## Loading the Software

**1** Insert the Final Cut Express CD into your computer.

**2** Double click the Final Cut Express Folder.

**3** Double click the Install Final Cut Express icon.

Follow the on-screen instructions – making sure you accept the license agreement.

**4** Check Boris Calligraphy and FXScript DVEs from CGM, unless you specifically do not want the extra functionality which these provide.

☒ Final Cut Express
☒ Boris Calligraphy
☒ FXScript DVE's from CGM

Make sure the program will be installed on the hard drive where your operating system is already installed. Relax while the computer runs through the installation process.

 Installation was successful. If you are finished, click Quit to leave the Installer. If you wish to perform additional installations, click Continue.

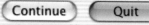

You will then be prompted that installation was successful.

**5** Quit the installation process unless you have other programs you wish to install.

## Initial Setup

Once Final Cut Express has been successfully installed on your computer you need to be able to access the application and set the program to the video standard you are working with.

# HANDS ON

The Final Cut Express application is found in the Applications folder which is located on the hard drive where the operating system for your computer is installed.

The easiest way to get to the applications folder is to choose the menu at the top of the desktop screen titled GO.

| Go | Window | Help |
|---|---|---|
| Back | | |
| Forward | | |
| 🖥 Computer | | |
| 🏠 Home | | |
| 💿 iDisk | | |
| **A: Applications** | | |

**1** Select the GO menu and scroll down to Applications.

**2** Locate the Final Cut Express application icon in the Applications folder.

📽 **Final Cut Express**

**3** Drag the Final Cut Express icon onto the dock and position it where you like.

**4** Click once on the Final Cut Express icon to launch the program. You will be prompted to enter your details, including serial number. Take care not to confuse zeros with the letter O or the letter I with the number 1. Make sure you enter details for all the required fields.

**5** Once your details are entered click the OK button which will pulsate in blue – press it and you're away.

When Final Cut Express opens for the first time you will be prompted to set the video format you are working with.

**1** There is a range of options to choose from. To keep things simple for the moment, select either DV-NTSC or DV-PAL.

```
DV - NTSC
DV - NTSC 32 kHz - FW
DV - NTSC 32 kHz Anam
DV - NTSC 32 kHz FW Ba
DV - NTSC 32kHz
DV - NTSC 48 kHz - FW
DV - NTSC 48 kHz Anam
DV - NTSC 48 kHz FW Ba
DV - PAL
DV - PAL 32 kHz
DV - PAL 32 kHz - FW Ba
DV - PAL 32 kHz Anamo
DV - PAL 32 kHz FW Bas
DV - PAL 48 kHz - FW Ba
DV - PAL 48 kHz Anamo
```

13

2. Click on the area labelled Primary Scratch Disk. Select the drive where you wish to store your video files. If you only have one drive you must work with that drive.

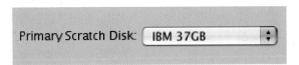

3. Click OK.

If you get a message saying 'unable to locate external video device' this means that no Firewire device is plugged into your Mac or that it is switched off. You need to check your Firewire device is plugged in properly and that it is switched on before you can go any further.

As technology progressed much of the post-production industry moved from film production to video production. Moviolas and flat-beds gave way to a new way of working: the two-machine video editing suite.

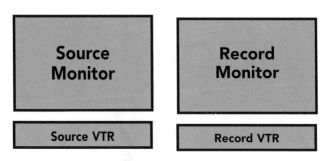

In this environment the editor would line up a shot on a source machine and edit across to a record machine. In and out points were marked, tapes were pre-rolled, run up to speed and images in the form of electronic signals were copied from one machine to the other.

The Final Cut Express interface is modelled on the same idea.

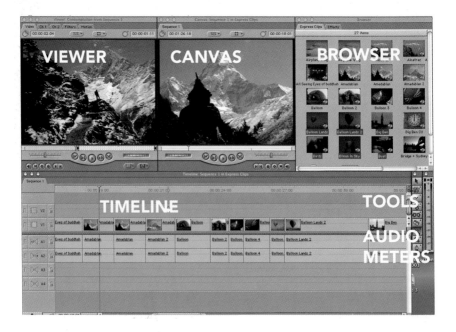

Notice the two windows located at the top left of the Final Cut Express interface. Think of the window on the left, called the Viewer, as being the source monitor and the window on the right, the Canvas, as being the record monitor. In essence, a shot is lined up in the Viewer and copied across to the Canvas. The area immediately below the Viewer and Canvas is known as the Timeline. This shows the edited shots as blocks in the order which they have been edited. The area window located top right above Timeline is called the Browser. Think of this as being like a cabinet which stores the masses of videotapes on shelves ready for the editor to access.

Take note of the Audio Meters and Tool Palette.

All professional VTRs have level meters which must be watched to make sure the audio doesn't distort during playback and transfer of sound and picture. The golden rule is always to make sure the audio meters do not peak into the red (DV audio should peak somewhere between −12 dB and −6 dB).

To swing the analogy back to the film days, the Tool Palette represents the tools the editor would physically work with: the splicer, the hand-winders, the spools and frame measuring instruments. The Tool Palette in Final Cut Express gives the editor access to the instruments with which the finer details of the editing process are crafted.

It should be clear by now that Final Cut draws on the very best the world of post-production has offered in the history of film and video production. Never before has the editor had better means to produce their creative vision.

If you find the terms Browser, Viewer, Canvas, and Timeline difficult to remember or to identify with, just think of the Browser as the area where all clips are stored, the Viewer is where one watches the individual video clips, the Canvas is where the material is edited and the Timeline represents the individual shots which make up the entire movie.

## Arranging the Interface

The Final Cut Express interface can be set up in several different ways. Individual users can work according to their own particular preference with regards to the layout of the interface. Several different arrangements can be chosen from within Final Cut Express or the editor can create up to two custom layouts.

1. Go to the Window menu (located top right) and scroll down to Arrange. You will notice there is a list of options for arranging the interface.

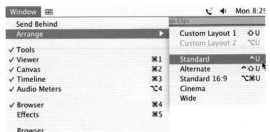

2. Scroll to any of these options and release your mouse button. Each time you wish to try a different layout you need to return to the Window menu, scroll to Arrange and then move across to the layout you wish to select.

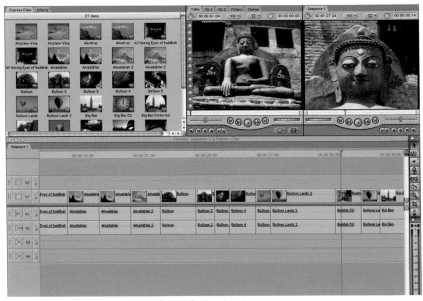

**The Wide Setting Layout**

I favor the Wide setting. This provides a large Timeline and presents the Browser, Viewer and Canvas at all roughly the same size and positioned next to each other.

THE INTERFACE

My preference is to modify this setup slightly. I move the Toolbar to the left, effectively positioning the Timeline central to the Toolbar and Audio Meters. This produces a neat, symmetrical display.

To achieve the above setup is simple:

1. Drag the Toolbar by clicking in the gray area at the top and position it on the opposite side of the screen beneath the Browser and next to the Timeline.

2. Slide the Timeline to the right so that it is positioned directly between the Toolbar and the Audio Meters.

3. If necessary resize the Timeline window by dragging the bottom right corner so there is no overlap onto either the Toolbar or Audio Meters.

Once you have set the layout according to your personal preference it is then possible to save the setup as a Custom Layout.

## Learning a Custom Layout

To set a custom layout it is important to execute the key/mouse combination as follows:

1. Hold down the Alt/Option key (located 2nd to left of the Space Bar).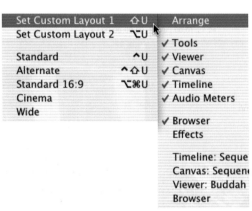

2. Whilst still holding down the Alt/Option key, select the Window menu at the top of the screen.

3. Scroll to Arrange. Where it normally displays Custom Layout 1 it will now read **Set Custom Layout 1**. Point your cursor to this setting and release the mouse button. Your Custom Layout will now be set.

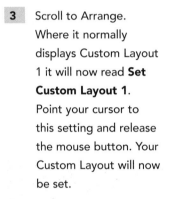

You can confirm this has been achieved by selecting any of the other layouts. Now go back to the Window menu, scroll to Arrange and select Custom Layout 1. Your screen should revert back to the Custom Layout you have just set. If it does not, backtrack using the instructions above and try again. Once your Custom Layout has been set it will be remembered each time you open up Final Cut Express and you can then choose your Custom Layout, or any of the setups which are listed under the arrange options.

It is possible to set up to two Custom Layouts, if you wish. This can be convenient when there is more than one editor that uses the same system or if you find different layouts suitable for different aspects of working within the program.

## Building a Film Structure

Any film is literally built. Just as a novel will have chapters and subplots, each and every film has an underlying structure. The raw components needed to build the film are planned for in the scripting stage, gathered while filming and structured during the editing.

I liken the film-making process to making a set of chopsticks from a tree-trunk. An entire tree can be whittled away to leave nothing remaining other than two small pieces of wood – these are the chopsticks. Film or video is the same. A mountain of footage is acquired and throughout the editing process this footage is chopped down to a fraction of its original size to leave a small, yet refined, remnant of the original content.

In the old days the film editor would take a piece of film, cut it and join another piece of film to it. The editor now had two shots in sequence. Another cut, another splice – three shots. More cuts, more edits – eventually a finished film was created.

When working with Final Cut Express exactly the same applies. No difference! Shots are cut together. Sequences are built. The end result is a finished film.

Look at the building blocks – each of these blocks represents a shot. By putting the blocks together, one after the other, a film is slowly built. Just like a house, one brick at a time.

Notice the blocks are different colors. Blue is for picture, green is for audio. The blocks, which represent the individual shots, can easily be reordered by simply sliding them around in the Timeline.

## Important Details About the Interface

It is worth having a good understanding of the interface of Final Cut Express. This gives you the power to use the program to its full potential and to achieve a variety of editing tasks in many different ways.

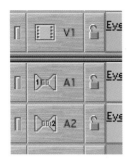
Look to the extreme left of the Timeline – notice there are green buttons next to each of the tracks. These are monitoring buttons for video and audio. If you press the green button on any of the tracks you are effectively switching it off – this will deactivate the monitoring for that particular track. This gives you the ability to mute the audio, or kill the video, at the flick of a switch.

At the bottom left of the Timeline there is a symbol that looks like two mountains – this is called Clip Overlays. Later, as you get into the editing and sound mixing process, you will find this facility extremely useful for adjusting audio levels and setting the opacity of video clips.

To the right of Clip Overlays are four little boxes. These boxes affect the size of the clips as they are displayed in the Timeline. This is useful for increasing the visual size of the clips if you are working with a monitor which is cramped for screen real estate.

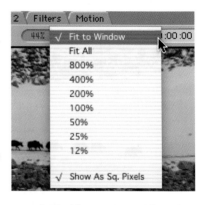
The following point is very important to take note of: look to the top of the Viewer and Canvas, just below the tabbed sections. You will notice there is a button with a percentage value in it. Click this button and it will reveal a series of numeric values – always keep this set to 'Fit To Window' and have 'Show as Square Pixels' checked at the bottom. If you do not select 'Fit To Window' you may encounter jerky playback and experience a great deal of frustration working out the solution. This applies to both the Viewer and the Canvas.

Considering that we have not even begun the editing process, the relevance of these details may seem a bit obscure at this stage. Rest assured it will make sense as you become familiar with the inner working of Final Cut Express.

## Saving Projects

To insure yourself against against loss of time and effort you need to save your project regularly. Final Cut Express will do this for you at pre-defined intervals (set in Preferences, found under the File menu).

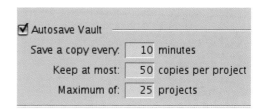

The files are saved into an area known as the Autosave Vault. I usually set this to save every 10 minutes and store a maximum of 50 projects. Should you run into any trouble along the way you can then backtrack and open an earlier version of your project.

The Autosave Vault is hidden away under your User Name in your Documents Folder. Here you will find the Autosave Vault inside the Final Cut Express Documents Folder.

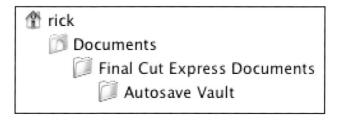

The easy way to find the Autosave Vault folder, without manually navigating through your computer, is to let the computer do the work for you. Click anywhere on the Mac desktop. Hold down the Apple key and press the letter 'F'. This will open the search dialog box. Type in the word Vault and select Everywhere from the option box at the top of the Search window. When the Autosave Vault is located click on the folder icon and this will open the Vault for you. Locate the project folder you are working with and inside of this will be the saved versions.

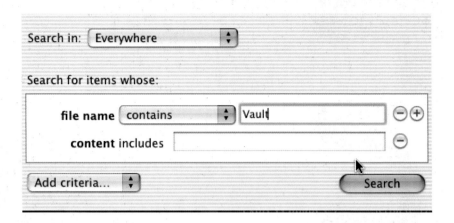

I usually save multiple versions of the same project straight to a folder of my choosing or to the desktop. This is easily achieved by selecting the Save Project As... command under the File menu in Final Cut Express. By saving the project regularly, with the time and date in the title, several versions are then stored on disk. In the event of a crash or a corrupt file, or user error, one can quickly return to an earlier version, or simply access the Autosave Vault. I call this double insurance.

# SETUP AND CAPTURE

By now you are probably itching to get going on the editing process. There are just a few more crucial areas to understand before the editing can begin.

Nothing can be done with any editing program if you do not have the video material stored on the hard disk of your computer. This process is known as Capture. Before you can capture your raw footage you must first set up your system. This is very easy to achieve. However, before you set up the system it is important to get a few technical details worked out. Set up your system incorrectly and you will have more trouble than you wish to think about. Get it right and your workflow will be trouble-free and let you concentrate on the creative aspects of the editing process.

## DV Audio

DV audio can be recorded at two different sample rates:

**16 Bit – 48 kHz**
**12 Bit – 32 kHz**

16 Bit – 48 kHz provides the highest quality and allows for two channels of audio, or a single stereo pair to be recorded.

12 Bit – 32 kHz provides lesser quality, though still very good, and allows for two sets of stereo pairs, or four individual tracks to be recorded.

For the most, 16 Bit audio is the preferred option, unless one specifically needs to access four independent channels. While this may sound ideal no DV cameras actually have inputs to record four independent channels of audio. The main advantage to setting a camera to 12 Bit – 32 kHz is that audio dubbing can then take place onto the remaining free set of stereo pairs.

Unless you specifically plan on accessing these tracks I recommend setting your camera to 16 Bit – 48 kHz. This is usually accessed through the menu settings in your camera.

Just to throw a spanner in the works there is another sample rate that needs to be considered. CDs are recorded at 44.1 kHz. This will be discussed later when capturing audio from CDs is covered.

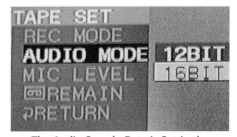

The Audio Sample Rate is Set in the Menu of Your DV Camera

## Easy Setup

Apple have made it very easy to set up Final Cut Express, however, it is up to you to make sure you get these settings right. If you set the audio sample rate incorrectly the result will be sync drift; if you set the video to widescreen anamorphic when it was shot in standard 4 × 3 your images will not fit the frame correctly.

To correctly set up Final Cut Express you need to access the Easy Setup menu.

1.  Open the Final Cut Express menu at the top left of the screen. Scroll to Easy Setup and release your mouse button.

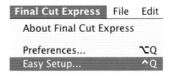

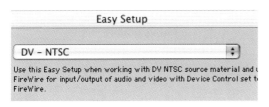

2.  Click with your mouse on the setup bar within Easy Setup and choose the relevant option. Most often DV-NTSC or DV-PAL is the appropriate setting. Both of these will automatically set the audio sample rate at the highest quality which is 48 kHz.

Avoid the Firewire Basic settings unless you find your camera or deck does not respond when you try to control the deck from within Final Cut Express. The reason Firewire Basic has been included is because different manufacturers use slightly different ways of implementing the Firewire standard. If you do encounter problems controlling your Firewire device then you will need to use one of the Firewire Basic settings.

Make sure you select a setting with 32 kHz if your material is recorded at the lower sample rate. Likewise, if you are filming in widescreen 16 × 9 aspect ratio you will need to choose one of the anamorphic settings.

## Setting Scratch Disks

You need to define an area where your captured video files will be stored – this area is called scratch disks. Final Cut Express provides the facility to set up to 12 scratch disks. While it is unlikely you will have 12 hard drives attached to your system the facility is there if you need it.

You already set a primary scratch disk when you loaded Final Cut Express onto your computer. Before you begin capturing you should check that the files will go to the drive or drives of your choosing. Setting the scratch disks is essential to working effectively with Final Cut Express.

To set the scratch disks you need to select Preferences which is found under the Final Cut Express menu at the top left of screen.

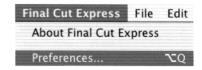

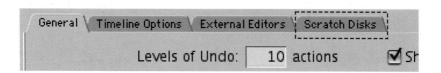

1. Open the Preferences window. Look to the series of tabs which are stacked from left to right at the top of the window.

2. Double click the Scratch Disks tab – located top right of the preferences window. This will reveal the Scratch Disks window.

3   Press the Set button closest to the top – the reason there are several Set buttons is to allow one to set multiple scratch disks.

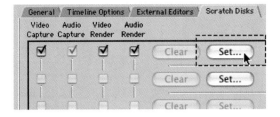

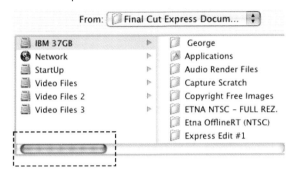

4   Using the blue bar at the bottom of the window scroll all the way to the left. This will show a list of all the available hard drives on your computer.

5   Select a hard drive where you want to store your video files by double clicking. This drive will then be set as the first scratch disk in the list. If possible, select a drive which does not contain the operating system for your Mac (this ensures optimum performance when editing).

6   Alternatively, you can single click a hard drive of your choice and then press the New Folder button. Name this folder, for example Media Files (or the name of your project), and press the Create button.

7   Press the Choose button and this will return you to the Scratch Disk window. Make sure you have Video Capture, Video Render and Audio Render all checked with a cross. By default this will already take place for the top Scratch Disk.

You have now defined an area to store your video files.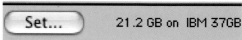

If you double clicked a drive at point 5 three folders will have been created on that drive – one for video files, one for video render files, and one for audio render files. If you moved on to point 6 you will have created a named folder inside of which three folders exist for video files and render files (for both video and audio).

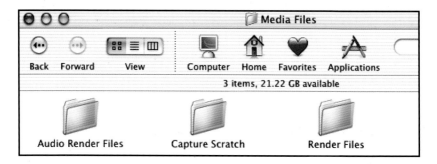

You can set up to 12 scratch disks depending on how many hard drives you have available. This is done by simply repeating the above procedure; however, select the Set buttons in order from top to bottom as you set each new Scratch Disk. It is an advantage to set several Scratch Disks as Final Cut Express automatically fills up the next available hard drive once the first hard drive becomes full.

Setting scratch disks will become second nature once you have run through the process a few times. The method described gives you the power to define exactly where you want your files to be stored. Providing you are disciplined you will easily be able to manage your media for each project you work on. When you come to the end of a project it will then be quite simple to get rid of the files you no longer need. Simply go to the area you defined when setting your scratch disks and delete the files which are no longer required.

## Working with Formats other than DV

Final Cut Express has been designed to capture material which originates on either the DV or DVCam format. In the real world Final Cut Express can work with ANY video format providing the original material is transferred to either DV or DVCam for editing. This is already an accepted way of working for those wishing to edit footage on a Firewire editing system which originates on formats outside of the DV spectrum.

Therefore should you have video footage which originates on another format, such as Beta SP, Digi-Beta, S-VHS – simply get your original footage transferred to either DV or DVCam. You can then capture the material into Final Cut Express using the same procedure used to capture footage which originated on DV. Most post-houses or dubbing facilities can arrange for this transfer to be done, although it can be cheaper to hire in a deck yourself and to do your own transfers, depending on which format/s and how much footage you need transferred in the first place.

## Methods of Capturing DV Footage

There are two ways to capture DV footage when using Final Cut Express: these are **Capture Clip** and **Capture Now**.

**Capture Clip**, as the name suggests, is used to capture a single clip at a time. It requires the editor first to mark 'in' and 'out' points. An 'in' point refers to the position on the tape where the capture process is to begin and the 'out' point is where the capture process is to stop. Once the 'in' and 'out' points are marked the computer cues up the tape in the deck/camera to the appropriate point and transfers the material onto hard drive.

**Capture Now** is used to capture clips 'on-the-fly'. This means the capture process begins the moment the editor instructs the computer to begin capturing and stops when the escape button is pressed.

## Deck Control

To capture video files to hard drive it is essential to know how to control the replay Deck or Camera from the computer. This is quite simple and has been well integrated into the editing interface. All operations are easily accessible using keyboard commands.

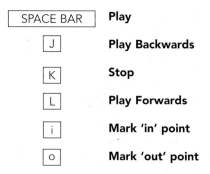

Each of the play commands – J and L – work in increments. By pressing J or L up to five times will speed up the result. This will be obvious as we get further in the Final Cut Express workflow.

## The Capture Window

The Capture window is the facility provided within Final Cut Express to enable you to perform the capture process. It is important to understand the controls within this window and how to use them.

1  To open the Capture window first make sure your deck/camera is switched on. If you are using a camera make sure it is in VTR mode.

2  Choose the File menu at the top left of the screen, scroll down to Capture and release your mouse button.

The Capture window will now open.

# SETUP AND CAPTURE

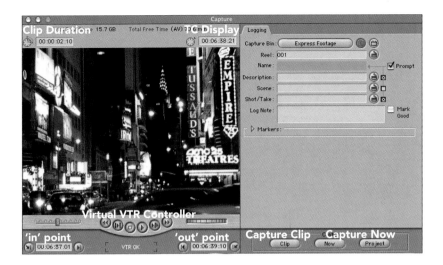

Above are marked the crucial areas one needs to understand to effectively use the Capture window.

**Clip Duration** – when using the Capture Clip setting an 'in' and 'out' point must first be marked. The duration of the clip is calculated by Final Cut Express and displayed in the Clip Duration window.

**Time Code Display** – whenever a DV tape is playing, a running display will show the timecode numbers ticking over. If you stop the tape the timecode at the exact point where the tape is parked will be shown.

**'In' Point** – an 'in' point is marked by pressing the letter 'i'. The marked 'in' point is displayed in this window.

**'Out' Point** – an 'out' point is marked by pressing the letter 'o'. The marked 'out' point is displayed in this window.

**Virtual VTR Controller** – just like most VTRs have stop, play and shuttle commands, this virtual controller performs similar functions.

**Capture Clip/Capture Now** – used to perform the capture functions.

Also note the display at the top of the Capture window which tells you how much free space is available on your computer and how much this capacity

33

equals in minutes. The amount of space is the sum total available on the scratch disk or disks you set earlier.

| Total Free Space 15.7 GB | Total Free Time (AV) 70.6 min |
|---|---|
| 00:00:02:10 | 00:06:38 |

You can therefore determine whether you have the room on your hard drives to capture the material required for your project.

## Capture Clip

If you wish to capture a single clip at a time this is easily achieved using the Capture Clip method. When using this method you need to first mark the 'in' and 'out' points for the clip you wish to capture.

1.  Put a DV tape into your deck or camera and make sure it is switched on. If you are using a camera make sure it is in VTR mode.

2.  Open the Capture window which is accessed through the File menu. Alternatively, press the Apple key and the number 8 and this will achieve the same result (the Apple key is located immediately left of the Space Bar).

3.  Press the Space Bar and your deck or camera will spring to life. If it doesn't, press play on your deck/camera to engage the heads. From this point on remote control of your Firewire device will work direct from the keyboard.

Once the tape is playing at speed, the result of having pressed the Space Bar, you can then spool through the tape using the J K L method. As mentioned earlier, pressing the letter J will run the tape backwards, K is for stop (or use the Space Bar to start/stop the tape) and the letter L is to run the tape forwards. Pressing the letters J and L many times affects the replay speed incrementally. If you press the letter J once the tape will spool backwards at normal speed, press it again and the tape will continue backwards, although slightly faster. Press it again and the speed will increase until the maximum speed is attained after five taps. Likewise, when using the letter L the tape will shuttle forward in increments until a maximum speed is achieved after five taps.

4   When you get to the point where you want the capture to begin press the letter 'i' – this will mark the 'in' point. Similarly, press the letter 'o' to mark the 'out' point. If you look to the bottom of the Capture window, the timecode reference for the marked 'in' and 'out' points will be displayed.

**Note:** when marking 'in' and 'out' points you can mark the points on-the-fly. That is by pressing the letters 'i' and 'o' at the appropriate point while the tape is running. If you prefer, while the tape is playing, hold down the letter 'i' and release it when you get to the point where you wish the 'in' point to be marked. The same applies for the 'out' point. Hold down the letter 'o' and release it to mark the 'out' point. Final Cut Express is flexible in that the same result can be achieved in a variety of ways.

5   Once the 'in' and 'out' points have been marked press the Clip button at the bottom of the Capture window. You will then be prompted to name the clip.

6   Press OK and the Mac will instruct the deck/camera to cue up the clip which will then be captured to disk and placed into the Browser for you to access.

By repeating this process you can capture as many clips as you wish.

## Capture Now

An alternative way to capture clips is to use Capture Now. This is a simple method that does not require you first to mark the 'in' and 'out' points.

1   Open the Capture window and play the tape in your deck/camera.

2. Press the Now button which sits immediately to the right of the Clip button in the Capture window. Immediately upon pressing 'Now'  the capture process will begin. The images on the DV tape will be mirrored in a large window on your computer monitor. A message at the bottom of this window will confirm that capture is taking place.

> Capturing Clip – NOW CAPTURING (press 'esc' to stop)
> WARNING: Capture Now is limited to 30 min

3. Once the material you want has played, press the Escape key (located top left of your keyboard) to exit Capture Now. The capture process will stop and the clip will be placed into the Browser.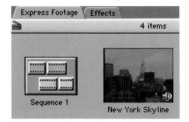

You can then name the clip by overtyping the name assigned to it by Final Cut Express.

Always remember to close the Capture window once you have completed the Capture process (do this by clicking at the extreme top left). Failure to do so will mean your video material will not play through the Firewire – to your Deck or Camera – and onto your Television Monitor (assuming you are working with this configuration). By closing the Capture window this problem will be avoided.

## Working Around the Lack of Batch Capture

The full version of Final Cut Pro provides a facility known as Batch Capture. This allows the editor to mark and log as many shots as one wishes and then to have Final Cut Pro cue up each of the shots and capture each of these individually. It is more efficient than capturing one shot, or a series of shots as a single clip, as is required with Final Cut Express.

Final Cut Express does not provide a Batch Capture facility. However, there is a work around. The work around is found by using a command known as Make Subclip.

Essentially a Subclip is a smaller part of a larger clip. It is therefore possible to capture a large section of video, using either the Capture Now or Capture Clip commands, and then to break this large clip into many smaller clips. While not as elegant as a true Batch Capture facility this does provide a way around the tedium of capturing many individual clips, one at a time. Simply capture a long piece of video, even an entire tape's worth, and then break this down into individual shots using the Make Subclip command.

The process of how to make Subclips will be discussed later in the editing section.

## Getting the Most Out of the Capture Process

An important part of preparing for the editing process is to watch your material. To edit your footage properly you have to know what is there to work with. I always use the capture stage as a dual purpose procedure: (1) to watch your footage, properly, in real-time as it is captured and (2) to capture the footage onto hard drive.

In the film world, as soon as the footage made it to and back from the lab, the dailies would then be screened. A projector was laced up and the illuminated images thrown forth in a display of light onto the screen. Based on what the director saw the rest of the shooting could then be arranged.

You are the director, the Log and Capture facility is your projector. So watch your images and build the structure of your film in your mind before you cut a single shot to another shot. You decide what to take in, what to leave out, and what to reshoot if necessary.

## Dealing with Timecode Breaks

When footage is recorded onto a DV tape a control track is recorded onto the tape. Written into this control track is a continuous series of numbers known as timecode. These numbers are used for many purposes during the editing, particularly in logging, when a person will perform what is called a pre-edit. In simple terms this means to sort through the raw footage and to perform a rough edit on paper before the footage is taken into the edit suite.

If this control track is broken, for any reason, the timecode numbers will not be continuous. Most often this is caused during shooting when the tape is stopped, perhaps to look at a shot, and then the recording begins again; however, after the point where the last take was recorded.

When tapes are brand new they are described as virgin tape. This means no control track has been recorded – the tape is completely blank. If, while shooting, the tape is stopped and perhaps taken out of the camera, and then reinserted to record more scenes, often a gap is left between the last section of recorded tape and the next section where the recording begins again.

The break in the control track is known as a timecode break. It is obvious this has taken place because the numbers on the tape are not continuous. Effectively what happens is the numbers continue to a point before being reset to zero, then to continue from that point forward.

The effect of a timecode break can cause loss of sync, a jump or break in the picture, and can be quite frustrating as the editor tries to work out what has caused these problems.

Should you experience a break in timecode, or any of the symptoms described above, then simply stop the tape and make sure you only capture footage prior to the timecode break. You can then resume the capture process after the timecode break. By being aware of these potential problems you can then work to avoid them.

## Importing Music from CD

Films are primarily made up of sound and picture. Much of the sound which accompanies a picture is recorded at the same time the picture is recorded. When it comes to music, more often than not, this will be sourced from a compact disc.

Importing tracks from CD into Final Cut Express is straightforward. To begin with you need to leave the Final Cut Express application and go to the desktop to access the CD.

**1**  Insert your CD into the Mac's CD drive.

**2**  Double click the CD icon on your desktop to open it.

**3**  Drag the track/s you wish to work with direct from the CD to the desktop – wait while the copy process takes place. You may wish to rename the CD track/s once the copy process is finished.

Once the track or tracks have been copied, you need to go back into Final Cut Express to import the tracks.

**1**  Make sure Final Cut Express is open in front of you.

**2**  Select the File menu and scroll down to Import. Scroll right and select Files.

**3**  Navigate to the desktop and locate the track/s you wish to import.

**4**  Highlight the track you wish to import. If you want to import more than one track hold down the Apple key and click each of the tracks with your mouse button.

**5**  Press the blue glowing Choose button.

The CD track will now appear in the Browser and is represented by a speaker icon. Rename the track if you choose.

An alternative way to import files into Final Cut Express is to drag them direct from the desktop into the Browser. This will achieve exactly the same result as using the import command.

## Converting Audio Sample Rates

It is easy to import CD tracks into Final Cut Express – the complicated part of the process is to get the CD sample rate to match that of the rest of your project.

You will remember DV audio is recorded at either 16 Bit – 48 kHz or 12 Bit – 32 kHz. The key to trouble-free audio editing within Final Cut Express is to make sure that all audio is of the same sample rate. Commercial CDs are recorded at 44.1 kHz. It is therefore advisable to convert the sample rate of the CD track to match the rest of the audio in your project.

**1** Highlight the CD audio track which needs to be converted in the Browser.

**2** Select the File menu, scroll to Export and select QuickTime.

**3** Click on the format bar to reveal a list of options.

**4** Select Aiff.

**5** Click the Options button which will reveal a series of settings. Where it reads Rate – click the two arrows facing in opposite directions. This will reveal a list of audio sample rates.

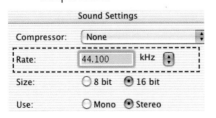

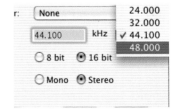

**6** Set the sample rate to 48 kHz and check that 16 bit and Stereo are selected. Click OK (if you are working with 12 bit 32 kHz audio you need to select the 32 kHz setting).

**7** Name the file and save it to hard disk. The conversion process will now take place.

**8** Go to the File menu and select Import. Locate the CD file you just exported and import this into the Browser. The sample rate has now been converted to 48 kHz and will now match the rest of your project.

What you have just done is to bring the CD track back into Final Cut Express at the same sample rate which matches the rest of your project. It is important to do this, otherwise problems such as pops and clicks at edit points and even loss of sync may occur. These problems are completely avoided by going through the process described above.

# SORTING THROUGH YOUR FOOTAGE

## Viewing Clips

Now that you have Captured your clips you need to be able to view them. This is the first step towards sorting through your footage. To view your material double click any of the clips in the Browser. Immediately the clip will open in the Viewer. Press the Space Bar and the clip will play.

The controls in the Viewer window can also be used to play the clip.

You can move quickly through the clip using the yellow Scrubber Bar, located below the image in the Viewer. Simply click once with the mouse and move the Scrubber Bar backwards or forwards.

It is also possible to shuttle through the clip using the J K L method. By tapping the 'J' or 'L' key up to five times the speed will **increase** or **decrease** in increments. Press 'K' or the Space Bar to stop. Press the Space Bar again to play.

To jog through the clip a frame at a time press the horizontal arrow keys located to the right of the Space Bar. The left arrow takes you backwards a frame at a time whilst the

right arrow takes you forward a frame at a time. Hold down the shift key and press either of the arrow keys and this will move forwards or backwards through the clip a second at a time.

## Playing Video Through Firewire

It is desirable to play the video signal through a deck or camera and onto a standard television screen. This is because the images on a television screen provide a true representation of the final quality of your finished movie. Otherwise you will be working exclusively off the computer monitor which provides a different type of picture to that of a television.

The deck or camera must be set to receive video through the Firewire cable. When using a deck you must check that the correct input is selected; when using a camera make sure the camera is switched to VTR mode. Beyond this the output from the Deck or Camera needs to be fed into the television set. Within Final Cut Express you need to select video to play through Firewire.

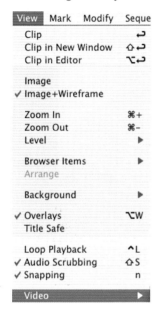

1. Choose the View menu found at the top center of the screen.

2. Scroll to video and select the Firewire option.

When using Final Cut Express, in order to access the Real-Time effects, one needs to edit exclusively on the Macintosh monitor. In this case one needs to select Real-Time from the View menu.

Make sure you have the correct type of Firewire cable to connect your camera or deck to your computer. You will need a large 6-pin to small 4-pin Firewire cable to do this.

## Setting Poster Frames

By default, when clips are captured in Final Cut Express they are displayed as miniature pictures, or icons. This way of representing clips is perfect for identifying a clip visually. The picture used to represent the clip, called the Poster Frame, is the first frame of the captured clip.

A problem arises when the first frame of the captured clip is not representative of the shot. Look at the example below. The image on the left does nothing to represent a horse in New York's Central Park, whereas the image on the right says it all.

It is possible to set any image from within the clip to be the Poster Frame:

1. Open the clip into the Viewer and position the Scrubber Bar on the frame you wish to display.

2. Select the Mark menu at the top of the screen and scroll down to Set Poster Frame. Release the mouse button and the image on the thumbnail will now change to that which you have selected.

It is also possible to reset the Poster Frame to the first frame of the shot simply by selecting the Clear Poster Frame command which is also found under the Mark menu.

## DV Start/Stop Detection

Back in the old days the film editor would take the workprint when it returned from the lab and cut it into pieces. These pieces were individual shots or sequences of film. The problem with having a huge amount of film on a reel, which had not been broken down into shots, was that it was unwieldy and time consuming to work with. Imagine if the editor of 40 years ago had a machine which would do that part of the process for them.

Well you do have it. Inside of Final Cut Express is a remarkable feature which will break up long captured clips into individual shots. Each time the record button is pressed on a DV camera a signal is sent to tape – another signal is then recorded to tape when the recording stops. Final Cut Express recognizes this recorded signal and once footage has been captured you can run your footage through what is called Scene Detection. Effectively Scene Detection scans through your footage and breaks it up into individual shots.

**1** Highlight a clip or group of clips in the Browser (a group of clips can be selected by dragging a lasso with your mouse around several clips or by holding down the Apple key whilst clicking on the individual clips).

**2** Select the Mark menu and scroll down to DV Start/Stop Detection – release your mouse button.

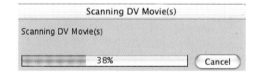

**3** A progress bar will appear as the clip or clips are scanned by the computer.

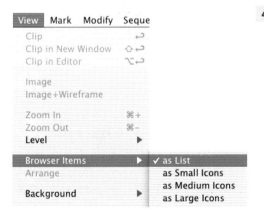

**4** Select Browser Items under the View menu – move right to select As List. The individual picture icons which represent each of your clips will now be displayed as a list of words rather than as pictures.

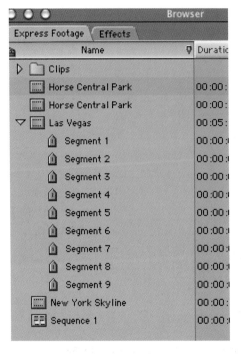

**5** Look to the clip you have just scanned. Notice there is a little arrow next to the clip name. Click the arrow and this will reveal a set of pink arrows. Each of these arrows represents a time when the DV camera has started and stopped recording. If you click any of the segments you can then view each of the shots in the Viewer.

DV Start/Stop Detection is a very useful function. If you wish you can capture an entire tape's worth of material and then have the computer break up the footage into individual shots. Effectively the computer does a great deal of sorting through your raw material for you. What it doesn't do is name the individual shots or sort out the good takes from the bad. That is something you must do.

## Working with Bins

In the old days before videotape was invented, and certainly before digital cameras and computers were used to acquire and edit productions, a film editor would organize strips of film in an area known as a Trim Bin. These film strips were hung on a horizontal rack and ordered according to the wishes of the film editor.

While a lot has changed technologically, when working with non-linear editing systems such as Final Cut Express, it is still crucial to order your material. Otherwise, it soon becomes impossible to track down your shots, particularly if you are working on a production with hours of footage and thousands of clips. Final Cut Express certainly has the power to handle productions of this magnitude!

# SORTING THROUGH YOUR FOOTAGE

To facilitate a simple way of ordering your material it is possible to create what are called bins within the Browser window. Within each of these bins you can store individual clips. The term bin, as you may have guessed, is taken from the era of film editing.

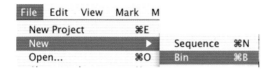

**1**   To create a new bin select the File menu at the top left of the screen, scroll down to New and select Bin. Alternatively, press 'Apple B' (the Apple button is located immediately left of the Space Bar).

**2**   In the Browser a box will appear titled Bin 1. This box is clear and different in shape to the clips so there is little likelihood of confusion. You can rename the bin by typing a name immediately after it has been created. Should you wish to rename the bin later, simply click once on the text area, then pause, and click again in the text area. You can now overtype the title and name the bin whatever you wish. Press return once you have renamed the bin.

Now that you have created and named a bin you can place clips inside it. Clips can be moved, one at a time, by clicking once on the clip and dragging it into a bin with the mouse. To select multiple clips use the mouse to drag a lasso around the clips you wish to highlight. Drag the highlighted clips over a bin and release the mouse button. The clips will then be dropped inside the bin.

**Highlight Clips and Drag into Bin**

Several clips can also be selected, one at a time, by holding down the Alt/Option key (located second to the left of the Space Bar) and clicking on each of the clips you wish to highlight. Drag the highlighted clips into the bin and release your mouse button.

49

To view the contents of a bin simply double click the bin and a floating window will appear with the contents visible in front of you. To order the clips in the window select the View menu and scroll to arrange.

**Floating Window**     **Select Arrange to Order the Picture Icons**

Once arranged, the result will be obvious. To close the floating window click the button at the top left and the bin will return to its original position within the Browser.

You can create as many bins as you like. You can also store bins within bins. Simply drag a bin over another bin and release your mouse button. The result is a bin stored within a bin.

Items can be moved from one bin to another by highlighting and dragging. Should you wish to move an item from within a bin back to the Browser, you must double click the bin to open it as a floating window and then drag the item or items out of the bin and into the Browser where they will be positioned.

If you want to delete either a clip or a bin highlight the item and press the delete key. Note that the items are only deleted from the Browser and not from your hard drive. Everything inside of Final Cut Express works by referencing to the original files which exist in the scratch disk folder/s which you set up earlier. Original clips remain stored on the hard disk of your Mac unless you actually go into the hard drive, remove the items and then place them in the trash on the dock. By emptying the trash the items are then deleted.

SORTING THROUGH YOUR FOOTAGE

Should you delete a clip from the Browser and wish to retrieve it, you can either import the clip from the hard drive where it was stored during the capture process – or you can locate the clip on your hard drive and drag it back into the Browser. Nothing is lost unless the original clip has been deleted from the hard drive where it was originally recorded.

## Working in List Mode

By default Final Cut Express will display clips as picture icons. It can be advantageous to view your clips in list mode, particularly when you are dealing with a lot of material. It can be much easier to scroll through alphabetical lists of clips displayed as words rather than searching through hundreds of pictures.

To view clips in List Mode:

1. Click on the Browser to make the Browser active.

2. Select the View menu at the top of the screen and scroll down to Browser Items.

3. Choose – as List. The items in the Browser will now be represented by words, rather than by icons.

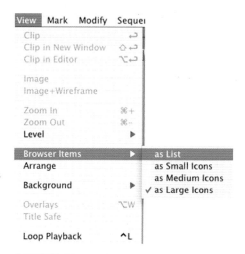

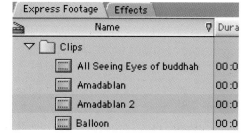

4. By clicking the triangle to the left of the bin's title the contents will be displayed in descending order.

My preference is to keep the items in the Browser in List View and items inside of bins in Picture View. By using this combination one has the advantage of being able to view the material both ways. If one clicks on the arrow to the left of any of the bins, the clips will be displayed in list view; whereas if one double clicks a bin the contents can then be displayed as picture icons.

If you have icons inside of bins, you may wish to repeat the process and, therefore, get everything into list mode. If you start your project in list mode, then items inside of the bins will also be displayed in list mode as you create them.

## Searching for Clips

Final Cut Express is a powerful editor capable of referencing to thousands of clips stored on the hard drives of your computer. As the editor you have to know what footage is there and how to get to it. It is all very well to know that it is there somewhere – if you can't find it you are lost… .

The most useful database of all is the human mind. An editor will constantly refer to the list in their mind to retrieve a shot ephemerally before actually doing so electronically.

When working on a large project, with hundreds or thousands of separate clips, you need a system to find what is needed. Providing you have taken care to label each of your clips in a way that is easy for you to identify with, you will then be able to search for any clip in your project. Apple has made this possible in an extremely simple and elegant way.

To search for a clip:

1. Click once anywhere in the gray area of the Browser.
2. Hold down the Apple key and press the letter 'F.' This will open the Search/Find dialog box. Alternatively, this can be accessed from the Edit menu by scrolling to the Find command.

## SORTING THROUGH YOUR FOOTAGE

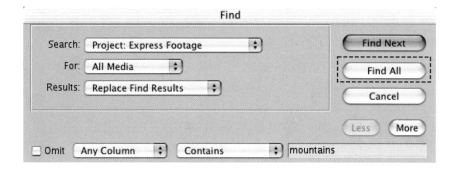

**3** Type the name of a clip or part of the name you wish to search for and then press Find All. All relevant items will then be placed in a bin as picture icons in front of you.

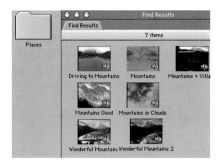

What would have taken a lot of effort in the past when working with film or tape has now been reduced to only a few seconds' work. You still need to rely on your mind to know what you are looking for. However, the difficult work of actually locating the clips has been made very easy providing you have labelled your clips correctly and you know what you are looking for in the first place.

The Search facility within Final Cut Express can be quite sophisticated. There are many options available beyond the simple method of searching within the Browser as just described. It is possible to search through all open projects, used and unused media, and the individual columns within the Browser.

Any Column
Name
Duration
In
Out
Tracks
Good (Y/N)
Log Note
Audio
Frame Size
Compressor
Aud Rate
Aud Format
Reverse Alpha (Y/N)
Anamorphic (Y/N)
Scene
Shot/Take
Reel
Comment 1

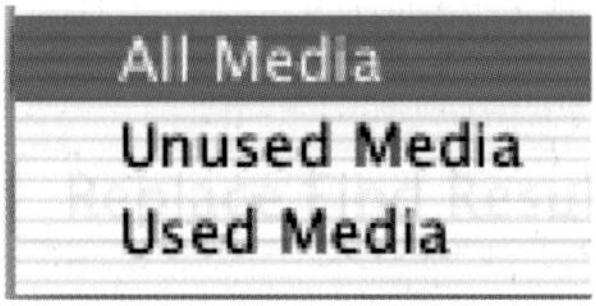

All Media
Unused Media
Used Media

All Open Projects
Effects
Project: Express Footage

Some of the possible search options within Final Cut Express

In simple terms editing is nothing more than putting shots and sounds together. In reality it is much more than this. It is both a technical and a creative process – it is also intuitive. Anyone can string words together, but not everyone is able to write a good story or a good book without a sound knowledge of language. Editing is similar … .

There are several key methods of editing with Final Cut Express. Of these the most important to understand are: Insert Editing and Overwrite Editing. While there are other methods available such as Replace Editing, Fit to Fill and Superimpose, provided you understand Insert and Overwrite editing you will be able to edit any production.

Let me stress, the key to editing with Final Cut Express lies in the difference between Insert and Overwrite Editing and when to use one or the other. Beyond this, you must understand how to control audio in relation to these two ways of editing. Once you have this clear in your mind you are well on the way to mastering Final Cut Express. You will then have the technical knowledge to make a film that is fully professional and equal to whatever you watch on TV or see at the cinema. I'm not kidding here. You will be able to edit anything from a thirty second commercial to a feature film.

## Insert and Overwrite Editing

Think of the old days when film was edited in a cutting room. The editor would take two pieces of film, line them up in a splicer and join them together. As more pieces of film were cut together a Sequence was formed. As more Sequences were joined these came together to build completed scenes until finally titles and effects were added. Once all the scenes were completed the final result was a finished film.

When putting the pieces of film together the editor had two choices: either, a piece of film was added to the shots already cut together and therefore the overall length of the Sequence was increased; or, a piece of film was placed into the Sequence and a corresponding amount of film, the same in length, removed – thus the overall duration did not change. These two choices are what **Insert** and **Overwrite** editing are all about.

When you build your movie in Final Cut Express you edit various shots together. Whenever these shots are put together you must decide whether you are adding a shot to a Sequence and therefore increasing the overall length of the movie, or, whether you wish to simply replace a section with another shot previously not included (thus keeping the Sequence length the same).

When editing with a non-linear system such as Final Cut Express the editor has a lot more in common with the film editors of yesterday than the tape editors of recent times.

## Getting Started with Editing

1. Check that you have a Sequence open. If you can see the Timeline in front of you then a Sequence is already open. Your Sequences are stored in the Browser, the same area where your clips and bins are kept. If a Sequence is not open, double click a Sequence in the Browser and the Timeline will appear.

2. Choose a clip from the Browser and double click it – this will load the clip into the Viewer (you may have to open one of your bins if you have filed away all of your clips).

3. With your clip loaded in the Viewer scrub through it. You can do this either by using the Scrubber Bar, or by using 'J' to scrub backwards or 'L' to scrub forwards (tap either of these keys in increments to speed up the rate of scrubbing). Use the Space Bar to start or stop.

4. Choose a point in the clip where you wish to mark an edit point. Press the 'i' key to mark the 'in' point.

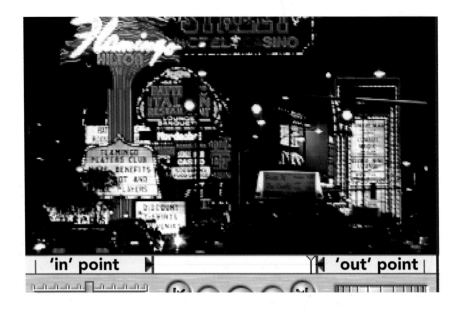

**5** Choose the point where you wish to mark the end point of the clip – this will be the out point. Press 'o' to mark the out point.

The procedure for marking 'in' and 'out' points is the same as that you already experienced during the logging process.

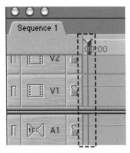

Scrubber Bar

Once you have marked the 'in' and 'out' points you are now ready to edit the shot into the Sequence. There should be no shots in the Sequence at this stage and therefore the Timeline will be empty.

Make sure the Scrubber Bar is positioned at the beginning of the Timeline. To do this click anywhere near the numbers in the light shade of gray at the top of the Timeline. You will now see the yellow Scrubber Bar with a vertical line extending from top to bottom. Drag the Scrubber Bar along this light gray area and position it at the beginning of the Sequence (all the way to the left). Alternatively, press the Home key on your keyboard and this will have the same effect.

**6** Click with your mouse in the center of the clip that you have loaded into the Viewer. A small transparent box will appear where you click. Whilst still holding the mouse button, drag the cursor over the Canvas (the window to the right of the Viewer). A selection of options will appear. The top option is Insert, followed by Overwrite, then Replace, Fit to Fill and Superimpose. At this stage we are only concerned with the first two options: **Insert** and **Overwrite**.

**7** Move the cursor, with the transparent box, over the Insert button (marked yellow). Release your mouse button. Look to the Timeline and notice there is now a single block positioned at the beginning. This is the first shot of your Sequence.

**8** Repeat the above process with another shot. Double click a shot to load it into the Viewer, mark the 'in' and 'out' points. Click on the shot in the Viewer, drag this over the Canvas and release it over the yellow Insert Button. You now have two shots in the Timeline.

9  Edit several more shots together – choose between five and ten shots. When you have cut these together, use the Scrubber Bar in the Timeline to move back and forth through the Sequence. Position the Scrubber Bar at the beginning of the Sequence and press the Space Bar. The shots will play in the Canvas and onto your television monitor – assuming you are plugged into a deck/camera and video through Firewire is selected under the View menu.

**Note:** you can also scrub through your Sequence in the Canvas by using the Scrubber Bar at the bottom of the Canvas. The Canvas and the Timeline are linked in that the Timeline is a graphical representation of all the shots edited together in the Canvas. The Timeline shows individual clips as blocks, whereas the Canvas shows the shots as moving images.

## Distinguishing Between Insert and Overwrite

In the Timeline you should now have several shots edited together. Position the Scrubber Bar in the Timeline at the beginning of the Sequence. Press the upward arrow on your keyboard (located to the right of the Space Bar) and you will find you are now able to skip forward between each of the shots. Press the downward arrow and you will find you can skip backwards through your shots, one by one.

 Skip Forwards     Skip Backwards

Now, position the Scrubber Bar in the middle of the Sequence.

1  Open a shot in the Viewer and mark the 'in' and 'out' points.

2  Drag this shot over to the Canvas. However, this time, instead of releasing it over the Insert button, position it over the Overwrite button (marked red). Now release your mouse button.

3  The shot will be edited into the Timeline – but it will not push all of the other shots in front of it further along in the Sequence. Instead, it will

write over a portion of the Sequence beginning where your Scrubber Bar is positioned.

If it is not obvious that this has happened it may be necessary to condense the overall spread of the shots on the Timeline. To do this, look to the bottom of the Timeline and find the slider bar with two ribbed ends. Drag either of these ribbed ends and you will see that the Timeline can be expanded or contracted. This does not affect the length of your movie in any way. What it does is to affect the display of your Sequence.

Ribbed End

**Expanded View of Timeline**

**Contracted View of Timeline**

This is very useful when you have a Sequence that is long and you wish to be able to view the entire Sequence on the screen in front of you. It is also useful when you wish to expand the Sequence for fine control to allow precise positioning of the Scrubber Bar.

To make completely clear the difference between Insert and Overwrite editing it is advisable to condense the Timeline so the entire contents are visible on screen. You will then be able to determine the type of edit: if the Timeline has been made longer, you have performed an Insert Edit; if the length does not change you have performed an Overwrite Edit.

To be able to see the difference between Insert and Overwrite editing:

**1**  Position the Scrubber Bar on a shot change in the middle of the Timeline.

**2**  Open a shot in the Viewer and mark an 'in' and 'out' point.

**3**  Drag the shot from the Viewer to the Canvas and position it over the Insert button. Observe the Timeline as you release the mouse button and notice all other shots get pushed further along the Sequence.

**4**  Hold down the Apple key and press the letter 'Z'. This will undo the action you have just performed.

**5**  Repeat the procedure of dragging the shot from the Viewer to the Canvas. However, this time, release it over the Overwrite button. It should be apparent that a different effect has taken place. The shots in the Sequence are not pushed further along the Timeline – they all stay in exactly the same position. What has happened is that the shot you have just edited into

the Sequence has written over a portion of the Timeline. The length is determined by the 'in' and 'out' points in the Viewer.

If you look at the top left of the Viewer you can see the duration of the shot you are working with. This is measured in seconds and frames. If you change the position of either the 'in' or 'out' points Final Cut will calculate the new duration.

**Note:** it is not necessary to drag the video from the Viewer to the Canvas to perform an Insert or Overwrite Edit. If you prefer, mark the 'in' and 'out' points in the Viewer and use the F9 key for Insert editing and F10 for Overwrite editing. Using a single keystroke is often the most efficient way to edit.

## Three Point Editing

So far we have only marked 'in' and 'out' points in the Viewer with the positioning of the Scrubber Bar determining where the Insert or Overwrite Edit will be edited in the Timeline. It is also possible to enter the 'in' and 'out' points directly into the Timeline. Simply position the Scrubber Bar where you want to mark the 'in' point and press 'i' and similarly press 'o' where you want to mark the 'out' point'.

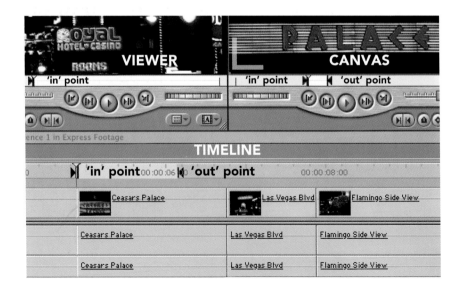

By marking a single 'in' point in the Viewer you can then perform an Insert or Overwrite Edit. The positioning and duration of the edit is determined by the 'in' and 'out' points marked in the Timeline.

It is also possible to mark the 'in' and 'out' points in the Canvas.

What is being illustrated here is known as Three Point Editing. Essentially, all editing in Final Cut Express works according to the Three Point Editing system. Even if it appears that only two points have been marked the positioning of the Scrubber Bar in the Timeline serves as the third point.

It is important to be aware that whatever points are marked in the Timeline will be reflected in the Canvas and vice versa. The Timeline and Canvas are intimately related at all times – they are in no way independent of each other.

It may be clear at this stage just how closely the Viewer and Canvas mimic a traditional two-machine editing suite. If one forgets about the Timeline for the moment, all that is taking place is marking 'in' and 'out' points in the Viewer and/or Canvas. This is the same process as marking 'in' and 'out' points in a two-machine edit suite with a Source VTR and a Record VTR.

## Other Editing Options

So far we have looked at Insert and Overwrite editing. You will have noticed other options can be chosen when one drags a clip from the Viewer to the Canvas.

**Replace Editing** – This is used to overwrite a shot into the Timeline, from the Viewer, with the duration being determined by the shot which already exists in the Timeline. By marking an 'in' point in both the Viewer and Timeline/Canvas, the shot being edited will match the duration of the shot being replaced in the Timeline.

**Fit to Fill** – Four points need to be marked to achieve a Fit to Fill edit: an 'in' and 'out' point in the Viewer and an 'in' and 'out' point in the Timeline or Canvas. The shot in the Viewer will then be either sped up or slowed down to fit into the space of the shot which is being overwritten in the Timeline. The overwritten section of the Timeline will then need to be rendered (dealt with later).

**Superimpose** – This is used for creating a second layer of video. When using this type of edit a shot is edited from the Viewer to the second video track in the Timeline. No apparent result will be noticed, other than that of an Overwrite Edit, unless the edited shot is reduced in size, effectively creating a picture in picture.

## Modifying 'In' and 'Out' Points

If you wish to clear 'in' or 'out' points there are several ways to achieve this.

1   Select the Mark menu at the top of the screen.

2   Scroll down and choose the relevant option: clear 'in' and 'out', clear 'in' or clear 'out'.

Keyboard shortcuts can be used to perform these functions. Hold down the Alt/Option button (two keys to the left of the Space Bar).

Alt/Option + x   **Clear 'in' and 'out'**
Alt/Option + i   **Clear 'in'**
Alt/Option + o   **Clear 'out'**

By holding down the Control key and clicking in the area where one scrolls with the Scrubber Bar, in either the Viewer, Canvas or Timeline, a contextual menu will appear. 'In' and 'out' points can be set or cleared by choosing the relevant command.

It is also possible to alter the 'in' or 'out' points by dragging or repositioning:

1   Click on the 'in' or 'out' point symbol in the Viewer, Canvas or Timeline and drag it to where you want it to be repositioned.

2   Alternatively, position the Scrubber Bar where you want the 'in' or 'out' point to be and press 'i' or 'o'. The 'in' or 'out' point is then repositioned and the previous 'in' or 'out' point is effectively deleted and the new one put in its place.

THE CUTTING ROOM

## Directing the Flow of Audio/Video

The editing that we have done so far has involved editing video and audio at the same time. To produce a professional film one needs to be able to edit video and audio independently. This is easy to achieve with Final Cut Express and as with many of the editing functions there is more than one way to go about it.

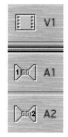

1. Look to the left of the Timeline and you will see to the left of the V1, A1 and A2 symbols are areas illuminated in yellow.

2. Click on the yellow area and it toggles between yellow and gray – Yellow means active whereas Gray is inactive.

Video Active Audio 1 and 2 Inactive

Video Inactive Audio 1 and 2 Active

A track is only active for editing if you have the yellow area lit. If, for example, you have V1 gray and A1 and A2 yellow, then only audio will be edited. The opposite applies if you gray out the yellow in A1 and A2 and leave V1 activated – then only video will be edited and the audio is not affected.

This is also illustrated when you drag a clip from the Viewer to the Canvas. The readout at the top of the Canvas will display the word 'none' next to the Track/s which are inactive.

## Locking Tracks

An alternative way to edit video and audio independently of each other is by Locking or Unlocking tracks. This provides a very simple and effective way to prevent either audio and video, or a combination of both, flowing through to a particular track or series of tracks. It is as simple as locking the track or tracks that you do not wish to alter.

Look at the Timeline on the left-hand side and you will see there are two video tracks and four audio tracks. This is the default number of video and audio tracks which Final Cut Express provides you with when you launch the program.

To the immediate right of the V1, A1 and A2 symbols are little locks. Click on the locks and notice the track or tracks become grayed out.

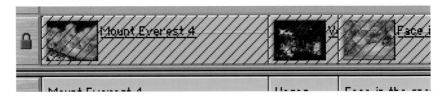

This means that when you edit material from the Viewer to the Canvas, and into the Timeline, the only part of the Timeline affected is that which is not locked (or not grayed out).

**Video 1 Locked**
**Audio 1 and 2 Unlocked**

**Video 1 Unlocked**
**Audio 1 and 2 Locked**

**Video 1 Locked**
**Audio 1 and 2 Locked**

To lock a track prevents it from being affected during editing. The only way to reactivate the track is to unlock it. This is done by clicking the lock on the left-hand side of the Timeline. Once the track is unlocked it no longer appears grayed out.

The usual rules regarding Insert and Overwrite Edits apply.

## Adding and Deleting Tracks

There are two ways to add or delete video and audio tracks to your project.

1. Select the Sequence Menu and scroll down to either Insert Tracks or Delete Tracks.

2. A menu will appear giving you the option to choose the number and type of tracks you wish to add or delete. You need to specify where these tracks are to appear in the Timeline by clicking the radio button options.

An alternative and easier method is to Control click in the gray area next to any of the tracks which already exist. A menu will open giving you the choice to either add or delete a track. This is the easiest method.

## Essential Editing Tools

A very important part of the interface that we have not dealt with so far is the Toolbar. With the layout I use this is positioned left of the Timeline. However, it really doesn't matter where it sits on the screen providing you can readily access it.

There are nine tools available. However, generally, I use only five of these for most editing tasks. There are also other tools hidden in the submenus within the Toolbar giving a total of 22 options in all.

**Pointer (Select Item)**

Edit selection

**Arrow (Select Track)**

Roll edit

Slip edit

**Razorblade**

**Magnifier**

Crop

**Pen Tool**

The five tools crucial to the editing process are highlighted to the left.

Mastering these tools is essential for producing films of professional standard. The other tools, while being useful to the advanced editor, are not necessary for most productions.

 **Pointer (Select Item)** – I call this the home tool. This is the tool I always have selected during the editing process. The Pointer is used for selecting and moving clips around in the Timeline. If I need to access the functions of the other tools I will choose another tool, use it, and then click on the Pointer again. By always having the Pointer selected you know where you are at all times.

**Arrow (Select Track)** – This is used for selecting individual or multiple Tracks, or the entire contents of the Timeline.

 **Razorblade** – Used for cutting clips into pieces. Great for trimming edits.

**Magnifying Glass** – Most useful for expanding and reducing the Timeline. Useful for homing in on the exact part of a clip you wish to work with.

 **Pen Tool** – Essential for adjusting audio levels. Also used for adding keyframes, thus allowing you to plot points over time. Useful for creating effects and adjusting video levels.

It is essential to understand how these tools work in order to edit efficiently. While shots can be strung together without ever touching the tools, in order to be able to trim edits, move shots around, home in on an exact part of a clip with absolute accuracy and to mix audio, one must be able to grasp these tools and how they should be used in combination with each other.

## Undo/Redo

You are now getting into the inner-workings of Final Cut Express. As your skills develop and you experiment with the facilities in front of you there will be times when you will get ahead of yourself and need to backtrack a few stages. This is easy to achieve in the form of undo. It is also useful to be able to redo any action you perform.

At any time an edit can be undone by holding down the **Apple key** and pressing '**Z**'. To perform a Redo command press **Apple 'Y'**.

**Apple + Z — Undo Action    Apple + Y — Redo Action**

Multiple levels of undo can be achieved by pressing Apple 'Z' again and again. Similarly, press Apple 'Y' several times to perform multiple redos. This ability to undo and redo is particularly useful when comparing changes in different edits. The number of levels of undo/redo can be set in Preferences found under the Final Cut Express menu located top left of your screen. The default amount is 10 levels of undo/redo. This can be set to a maximum of 99.

## Linked/Unlinked Selection

By default your audio and video are locked together. This means clips positioned in the Timeline will be married together in a similar way to film images and magnetic sound-striped tape running together in a synchronizer or a projector.

To illustrate the meaning of Linked Selection make sure you have selected the Pointer from the Toolbar. Check that you have several clips in your Timeline.

**1**  Point your cursor at a clip in the Timeline and click once. The clip is now highlighted.

**2** Whilst still holding down your mouse button, slide this clip to the right or left. Notice that both the audio and video move together.

**3** Release the clip you are moving over one of the clips in the Timeline and the video and audio will overwrite the clip over which it is positioned.

**4** Press Apple 'Z' and the Overwrite Edit will be undone. The clip will return to its previous position.

**5** Click once on the yellow symbol of two circles located top right of the Timeline. The yellow circles will turn white to represent that Linked Selection is turned off.

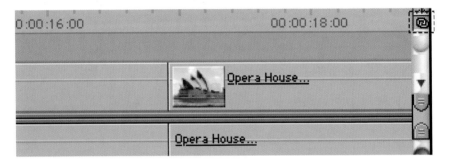

Linked and Unlinked Selection can also be accessed from the Sequence menu at the top of the screen.

Scroll down to Linked Selection; if there is a tick, this means Linked Selection is on; no tick and it is switched off. Linked and Unlinked Selection is turned on and off through the menu by means of a toggle.

With Linked Selection switched off repeat the procedure of selecting a video clip. Slide the video to the right and observe that while the video moves, the audio stays where it is. Conversely, select the audio and you can move this without affecting the video.

Video Moved Independent of Audio    Audio Moved Independent of Video

To select more than one track at a time hold down the Shift key, whilst using the Pointer tool, and items can be grouped together. Release the Shift key and the grouped items can be moved and repositioned wherever you wish.

To switch Linked Selection back on, click once on the white circles on the top right-hand side of the Timeline. These will turn yellow indicating that Linked Selection is switched on – this will apply even if you have moved video and/or audio independently of each other.

**Note:** an identical effect to linking or unlinking can be achieved by locking your tracks. Simply lock the tracks you do not wish to alter and then slide the video or audio of the clip you wish to move. Even though Linked Selection may still be turned on a locked track or tracks will over-ride the link.

## Moving Edits in the Timeline

You may have noticed that when you use the Pointer tool to slide a clip to a different location the effect is that of an Overwrite Edit. It is also possible to move edits around in the Timeline, using the Pointer tool, and at the same time perform an Insert Edit.

To perform an Insert Edit within the Timeline it is crucial to press the keys in the correct order.

1. Using the Pointer tool highlight the clip you wish to move and release your mouse button.

2. Press and hold down the Alt/Option key and click once again with the Pointer tool on the clip you wish to move. Reposition the clip by dragging and release your mouse button at the point where you want the clip to be inserted in the Timeline.

This time the result is that of an Insert Edit. The clip you have moved is repositioned and all edits in front of it move forward in the Timeline.

You may notice that the clip has been inserted where you specified in the Timeline and that it also remains in its original position.

To remove the original clip, highlight it and press the delete key. This deletes the clip from the Timeline and leaves a gap where it previously existed.

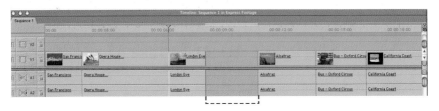

**Gap Where Clip Has Been Deleted**

To get rid of the gap hold down the Control key and click in the gap with your cursor – this opens a dialog box with several options – select Close Gap and the gap will disappear.

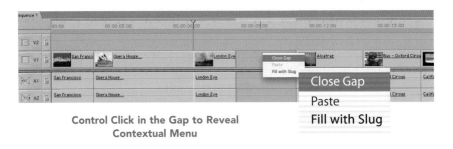

**Control Click in the Gap to Reveal Contextual Menu**

**The Gap Has Now Been Closed**

**Note:** you can actually delete a clip and close the gap at the same time by highlighting a clip – hold down the Shift key and at the same time press delete.

Most Apple keyboards have an additional Delete key to the right of the standard Delete key. Press this key and the close gap function is then performed with a single keystroke.

## Selecting Multiple Items in the Timeline

You will now be well aware that a clip in the Timeline can be selected by clicking once with the Pointer tool. If you wish to select more than one clip hold down the Shift key while you highlight each of the clips. So long as you continue to hold down the Shift key you can then select as many clips as you wish. These can then be moved within the Timeline using the Overwrite or Insert method.

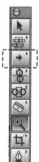

Another way to select multiple items is to use the Arrow tool. By clicking once and holding this tool can be extended to reveal various options.

Extended View of Arrow Tool

The Arrow tool is useful for deleting or copying large portions of the Timeline. Simply use the method described below to highlight those clips you wish to work with.

1   Click once to select the horizontal Arrow tool.
2   Your cursor now becomes a horizontal arrow. Use this horizontal arrow and click in the middle of the Timeline. All the clips forward of the arrow will now appear highlighted.
3   Press delete and the highlighted section will disappear.

**4** Press Apple 'Z' to undo the effect.

To copy the selected items repeat the procedure and this time instead of pressing Delete press Apple 'C' – or go to the Edit menu at the top of your screen and scroll down to Copy.

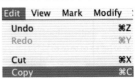

It is then possible to Paste these items anywhere in the Timeline.

**1** Select the Pointer tool.

**2** Place the Scrubber Bar at the position in the Timeline where you wish to Insert or Overwrite the shots you have just copied.

**3** Go to the Edit menu at the top of the screen. Select Paste to perform an Overwrite Edit on the section you have copied or select Paste Insert to perform an Insert Edit.

If you click in the Toolbar and hold the Horizontal Arrow down with your mouse the Toolbar will extend to reveal other options within the Arrow's capabilities. You can select a Horizontal Arrow which points backwards rather than forwards and thus select the contents of a track in the reverse direction; you can also select an arrow which points in both directions, allowing you to quickly and easily select the entire contents of an individual track or everything in the Timeline (providing Linked Selection is switched on). You can also select a double arrow, forwards or backwards, which has the effect of selecting all tracks in the direction of the arrows regardless of whether Linked Selection is switched on or off.

 Selects Track/s in a Forward Direction

 Selects Track/s in a Reverse Direction

 Selects Track/s in Both Directions

 Selects All Tracks in a Forward Direction Regardless of Whether Linked Selection is Switched On or Off

 Selects All Tracks in a Reverse Direction Regardless of Whether Linked Selection is Switched On or Off

## Edit Like Using a Word Processor

When using Final Cut Express shots can be Cut, Copied or Pasted using the conventions used in most word processors. These functions can be accessed from the Edit menu at the top of the screen.

To copy or paste a section from the Timeline is easy:

**1** Highlight one or more clips in the Timeline and Select Copy or Cut using the Edit menu at the top of your screen – or use the shortcuts Apple X (Cut) or Apple C (Copy).

**2** Go to the Edit menu at the top of your screen and select either Paste (Apple V) to perform an Overwrite Edit, or Paste Insert (Shift V) to perform an Insert Edit of the copied or cut material.

Paste Insert (Shift V) is an extremely useful function, not found in word processors. As described above, to perform an Insert Edit select Paste Insert whilst to perform an Overwrite Edit use Paste.

## Snapping and Skipping Between Shots

It is easy to skip between shots in the Timeline by dragging the Scrubber Bar which sticks to each of the Edit points. If you have a crowded Timeline you may wish to turn this facility off as it can make it difficult to position the Scrubber Bar with accuracy.

Press the letter 'N' and Snapping toggles on or off. You can also select the yellow button at the extreme top right of your Timeline to achieve the same result.

You can also skip between shots by using the vertical or horizontal arrows on your keyboard – up for forwards, down for backwards. Each press of these arrows will skip past one clip at a time.

## The Razorblade Tool

My favorite tool in Final Cut Express is the Razorblade. This tool is used for cutting clips into smaller pieces and is great for trimming a long shot into a smaller shot or shots.

It is often useful to use the Razorblade in conjunction with the Magnifier tool. By using the Magnifier you can zoom in on a clip or series of clips for greater accuracy when trimming with the Razorblade.

1   Play your Sequence in the Timeline. When you see a shot you would like to trim press the Space Bar to pause playback.

2   Click on the Razorblade tool – your cursor now becomes a Razorblade.

3   Position the Razorblade where you wish to cut the shot in the Timeline. The Razorblade will automatically be drawn to each edit point providing you have Snapping switched on. If you find the Snapping facility to be impeding your accuracy then switch it off. The Razorblade will also be drawn to the Scrubber Bar and it is often useful to position the Scrubber Bar at the exact point where you wish to perform a cut.

4   Click once to make a cut at the point where the Razorblade is positioned. The tracks to which the cut applies are determined by whether or not the tracks are linked.

**5** Select the Pointer tool and highlight the portion of the shot you wish to remove.

**6** Press Shift Delete or the Delete key to the right of the main keyboard area. This will remove the shot you have highlighted and close the resulting gap at the same time. If you wish to leave a gap only press Delete.

Should you happen to have Linked Selection switched off then the Razorblade will only cut through a single track at a time.

 If you wish to cut through all tracks, regardless of whether Linked Selection is on or off, then you need to select the Double Razorblade. This is done by holding your mouse button and clicking on the Razorblade tool – scroll across to select the Double Razorblade. The Double Razorblade will cut through all your tracks even if Linked Selection is switched off.

If you wish to heal a cut made by the Razorblade it is possible to perform what is called a Join Through Edit.

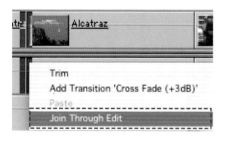

**1** Control click on the cut made by the Razorblade. This will bring up a contextual menu.

**2** Select Join Through Edit and release your mouse button. The cut will now be healed.

**Note:** the Razorblade does not cut through tracks which are locked. You need to switch off the locks for the Razorblade to work.

## Magnifier Tool

If you find it hard to be accurate when positioning the Scrubber Bar or Razorblade then you need to expand the Timeline. This is achieved by pulling on the ribbed ends of the Slider tool at the bottom of the Timeline, or by using the Magnifier tool.

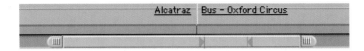

**Pull Either End to Expand or Contract the Timeline**

The main difference between the Magnifier tool and the Slider tool is that the Magnifier tool is used to zoom in on a specific section of the Timeline. By using this tool the exact area one wants to magnify will get larger with each press of the button. When using the Slider tool the overall spread of the Timeline is increased or reduced, but not necessarily the specific area you wish to focus in on. The Magnifier is far more accurate.

1. Click once on the Magnifier tool in the Toolbar – your cursor becomes a magnifying glass.

2. Position the Magnifying glass over the area of the Timeline you wish to enlarge. Click with the mouse and the Timeline will expand – click again and it expands further. You can continue expanding the Timeline until you are able to work with the individual frames.

3. To contract the Timeline when using the Magnifier press the Alt key and a minus symbol will appear, indicating that the Magnifier will contract the Timeline. Alternatively, click on the Magnifier in the Toolbar and select the minus Magnifier.

## Bringing Clips Back into Sync

Due to the nature of Insert editing and Overwrite editing, along with the fact that you can lock/unlock your tracks or activate/deactivate the yellow patching facility, it is inevitable that at some stage your video and audio will get out of sync. Fortunately, Apple has made it very easy to bring items back into sync.

If items are pushed out of sync a red box will appear at the beginning of the clip in the Timeline where the sync trouble has occurred. The red box, which appears in both the video and audio tracks, will have a figure indicating the amount of sync slippage that has occurred.

To bring the items back into sync:

1. Hold down the Control key and click inside the red box where the numeric value of sync drift is displayed. Control clicking in the red box brings up a dialog box with two possible options.

2. Select Slip into Sync or Move into Sync and release the mouse button. The item or items selected will now be brought back into sync.

It is also possible to manually slide the audio or video back into sync.

1. Make sure Linked Selection is turned off and Snapping is turned on. Having Snapping turned on will ensure that the out of sync items will be drawn to the correct sync points.

2. Highlight the item/s you wish to move and drag these so that the out of sync items line up at the beginning of each clip. When you release the mouse button the red box disappears and sync has been restored. If the red box is still visible move the clip again until you find the sync point.

## Creating New Sequences

Final Cut Express is particularly flexible in that you can have many Sequences open at a time. To have multiple Sequences open means that you have access to more than one Timeline – this is most useful for building various sections of a film which can be later joined together using the Cut, Copy, Paste and Paste Insert functions. Think of having multiple sequences as like having several different film reels each containing separate edited scenes or parts of a movie.

Multiple Sequences are tiled with cascading tabs from left to right. These tabs are displayed in both the Timeline and Canvas. Click on any of the tabs to flip between the Sequences. The Sequences can be renamed by overtyping the name in the Browser. The label on each of the tabs will then be updated.

1. To create a new Sequence click once inside the Browser. Hold the Apple key and press the letter 'N'. A new Sequence will then appear. Another way to create a new Sequence is to select the File menu (top left of screen). Scroll down to New Sequence and release the mouse button. A new Sequence will appear in the Browser.

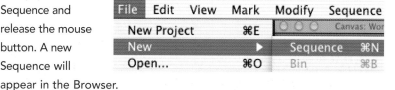

2. Double click on any Sequence in the Browser to open it. Several Sequences can be opened at any one time. It is possible to rename Sequences at any stage by highlighting the Sequence icon in the Browser and overtyping.

## Slow/Fast Motion

No editing program would be complete without being able to perform slow motion or fast motion to a clip or series of clips. Final Cut Express performs admirably in this area, giving the freedom to slow images down to 1% or to speed images up in excess of 1000%. It is also possible to play images in reverse.

**1** Click once in the Timeline and highlight the shot you wish to slow down or speed up.

**2** Go to the Modify menu at the top of the screen and scroll down to Speed.

**3** Enter a percentage value to set the speed you wish the shot to play at. If you want to play the clip at half speed enter 50%. If you want the shot to play at double speed enter 200%.

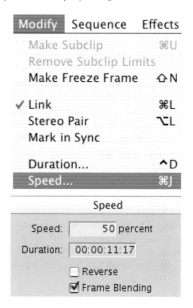

**Render Bar**

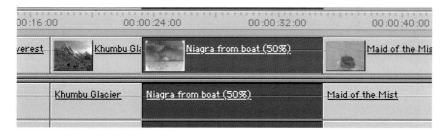

**4** The duration of the shot in the Timeline will now change according to the value entered. A red line will appear above the shot at the top of the Timeline alerting you that the shot must be rendered. (Rendering is dealt with in detail in the next section.)

5   Select the Sequence menu at the top of your screen and scroll down to Render Selection. (This will apply to the shot you have highlighted. If you have more than one shot to render then select Render All from the Sequence menu.)

6   Once your effect has been rendered you can play back the result.

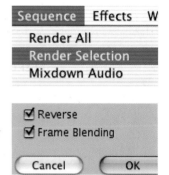

To play a clip in reverse simply check the 'reverse' option in the speed dialog box or enter a negative value.

It is worth noting when changing a clip's speed that the audio may be affected. It can therefore be useful to position the clip at the end of the Timeline, adjust the Speed and then Render to check the result. Once you are satisfied, the clip can then be repositioned in the Timeline wherever you wish.

## Rendering

Rendering is the process by which your computer builds each of the individual frames needed to produce an effect. When you play back straight cuts in the Timeline nothing needs to be rendered. The computer simply refers to the hard drive where the original shot has been recorded and uses your edit information to determine which section of the original shot is needed. When an effect is applied a different process must take place.

People often complain about rendering – to wait a few seconds for a computer to produce a one second dissolve can often drive people mad. I always smile at these situations. My background was in the world of on-line tape editing where the editor would work with several videotape machines, a separate vision mixer, character generator, audio mixer and Digital Video Effects (DVE) generator. To produce a dissolve, in this setup, would often require an editor to record one of the shots to a separate tape, thus dubbing the shot down a generation. The two shots would then be run together through a vision mixer where the dissolve would be mixed in real-time. Real-time was only achieved at the expense of the time used in the setup for the dissolve. To have to wait seconds for a dissolve would have been unimaginable.

However, in the real world many effects can take hours to render. Color correction, for example, where a great deal of footage must be processed, can be particularly time consuming.

Whenever you see a red bar at the top of the Timeline (above a shot to which an effect has been applied) this means this section must be rendered. It is possible to render individual clips or to instruct the computer to render everything in the Timeline that needs rendering.

**1**  When a shot requires rendering, go to the Sequence menu at the top of your screen and scroll down to Render Selection. Release the mouse button and the render will begin – a progress bar will display in percentage terms how much material has been rendered.

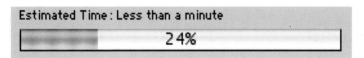

**2**  If you have more than one shot to render you can choose Render All from the Sequence menu and all the material to be rendered will be processed.

**3**  Once the render is complete you can then play the rendered clip in real-time. If you are happy with the result, carry on editing. If you are not happy make the necessary adjustments to the parameters of the effect and then render again.

At any time you can stop a render in mid-progress by pressing the Escape key (located top left of your keyboard). The render will cease; however, the portion of the shot or Sequence already rendered can be played back. This is a particularly useful feature as one can choose to render a small portion of a shot or Sequence, then play back this section to determine whether or not to go ahead with the complete render. Once you restart the render process begins again at the point where it was previously stopped. Thus, you do not need to re-render material that has already been rendered just because you stopped the process mid-render to look at the result of an effect.

## Using the Real-Time Effects

What is called Real-Time effects in Final Cut Express are really real-time previews – the results can only be seen on the computer monitor. To output to a deck or camera and onto a television screen you still need to render any effects.

The Real-Time effects are processor dependent which means the greater the speed of your processor, the more real-time you get. To access these effects you need a G4 processor with a minimum processor speed of 500 MHz or a twin-processor of 450 MHz or greater. The Real-Time effects cannot be accessed by any of the G3 machines.

To switch on the Real-Time effects capability you need to go to the View menu, scroll to Video and select Real-Time. Immediately the video signal will stop playing out of the Firewire, and all editing will now take place on the computer monitor inside of the Viewer and Canvas windows.

As you build your effects you will notice there are three types of colored bars which appear above the clips to which effects have been applied.

**Green** – indicates the effect will play in real-time.

**Yellow** – the effect will play in real-time; however, the result is an approximation of how the final effect will appear once rendered.

**Red** – the effect must be rendered to see the result.

Not everything works in real-time. None of the filters are real-time, though some of the transitions are. It is possible to do real-time picture in picture and two to several layers of video tracks (depending on your processor/s). Likewise, some titling and generator capabilities are possible in real-time.

So the Real-Time effects in Final Cut Express are at best a compromise – a compromise between working on the computer monitor, as opposed to seeing the images as they play through the Firewire onto a television. However, the real-time capabilities are definitely worth having – they can be particularly useful for working out timing decisions with fades, dissolves, motion effects and transitions. While one can complete a production to a high standard without Real-Time effects they are definitely desirable and provide a valuable tool for film-makers running machines capable of utilizing these effects.

## Subclips

One of the easiest and most useful functions available in Final Cut Express is the Make Subclip command. A subclip is a part of a larger clip – it is therefore possible to have a long clip and to break this clip into many smaller pieces which can be individually named. As mentioned earlier using subclips can provide a work-around for the lack of batch capture in Final Cut Express. Simply capture an entire tape's worth of material and then break the tape into many individual subclips.

1   Open a clip into the Viewer.
2   Mark an 'in' and 'out' point – this will define the beginning and end of the subclip.
3   Go to the Modify menu – select the Make Subclip command.
4   In the Browser a new icon will appear – the name will be that of the clip you have opened in the Viewer with the word subclip after the title.

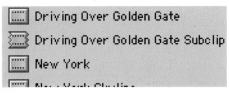

Driving Over Golden Gate Subclip

Once you have your subclips in the Browser you can organize them into Bins in the same way as you would with clips. The subclips can then be edited into the Timeline and worked with in the same way as clips.

## Freeze Frame

When working with NTSC video each second is made up of 29.97 individual frames; when working in PAL each second is made up of 25 frames. It is a simple procedure to freeze any of these frames and create what is known as a Freeze Frame.

**1** Position the Scrubber Bar on the frame you wish to freeze in either the Timeline or the Viewer.

**2** Select the Modify menu at the top of the screen and scroll down to Make Freeze Frame.

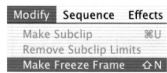

**3** Release your mouse button and the Freeze Frame with 'in' and 'out' points marked is positioned in the Viewer. By default a 12 second freeze is created (this can be set in Preferences under the Final Cut Express menu).

**4** If you want the Freeze Frame to be accessible within the Browser then drag the frame from the Viewer into the first column of the Browser. The Freeze Frame is represented by a graphic symbol within the Browser when viewed in List Mode.

THE CUTTING ROOM

Drag the Freeze Frame from the Viewer into the Browser

## Match Frame Editing

In on-line edit suites it is often necessary to perform a function known as Match Frame Editing. This means a specific frame is cued on a record machine, and an identical frame is cued on a source machine. By editing from the Source VTR to the Record VTR a seamless edit is performed. This was particularly useful in linear, multi-machine edit suites, before digital technology existed, when the editors did their best to minimize dubbing shots from one tape to another.

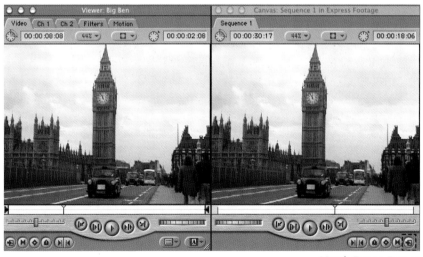

Match Frame Button

89

Match Frame Editing is still relevant in non-linear edit suites, such as Final Cut Express, although it can be used for different reasons. It can be extremely useful to be able to find an exact frame in a Sequence with absolute accuracy. This technique can be used to locate a clip quickly and to put this clip into the Viewer for easy access.

It is much easier to perform a Match Frame Edit when using Final Cut Express than it was in the on-line suites of yesterday. The editing process has been simplified thus allowing more time for creativity.

To achieve a Match Frame Edit:

**1** Place the Scrubber Bar in the Timeline on the frame you wish to match to in the Viewer.

**2** Press the 'F' key on the keyboard or press the Match Frame button in the Canvas.

**3** The frame on which you are positioned in the Timeline will now be displayed in both the Viewer and the Canvas.

**4** By marking the 'in' and 'out' points in the Viewer and (if required) the Timeline, you can perform a seamless frame accurate edit. This is what is known as Match Frame Editing.

While it may not be immediately clear exactly how useful this is, there are many instances where it can make the difference between being able to successfully create an effect or not. It is also the easiest method of finding a shot without having to look through the Browser and all of the Bins. Simply line up the Scrubber Bar in the Timeline – press the 'F' key – and the shot is immediately matched to.

THE CUTTING ROOM

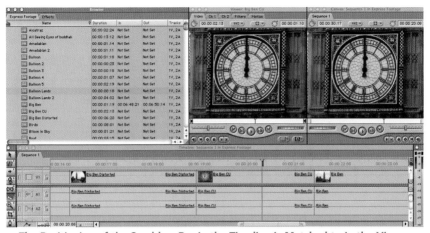

The Positioning of the Scrubber Bar in the Timeline is Matched to in the Viewer

## Split Edits

In most situations where an edit takes place both video and audio will cut at the same point. This is fine for most situations. However, in other situations a common technique known as a Split Edit may be used. A Split Edit is where audio and video are not cut at exactly the same place – either one may precede or proceed the other and this can also apply to either the 'in' or 'out' point of an edit – thus audio or video may start and/or finish at separate points.

Split Editing is a technique often used in news and documentaries, particularly in interview situations. For example, you may hear a person speaking over visuals of a scene being described. After several seconds, with the voice of the person still running, the video will cut to the person speaking.

To achieve a Split Edit:

**1** Open a shot into the Viewer.

**2** Place the Scrubber Bar where you wish to mark the edit.

**3** Hold down the Control key and click in the white area in the Viewer (where you mark 'in' and 'out' points). This will open the contextual menu which is used for clearing and setting 'in' and 'out' points. At the bottom of this menu is an option for Mark Split.

4. Scroll to Mark Split. This opens a menu to the right giving you options such as: Video In, Audio In, Video Out and Audio Out.

5. Choose an option, whether it be Video In or Audio In and release the mouse button.

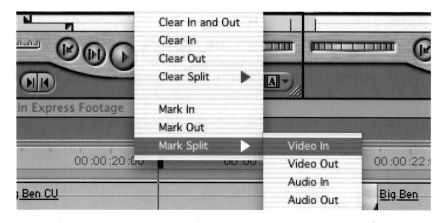

6. Reposition the Scrubber Bar and now choose the opposite choice – if you have already marked Video In, now choose Audio In.

7. Repeat the process for the end of the edit. You can mark separate Video and Audio Out points. If you do not want to separate Video and Audio Out points then mark an out point by simply pressing 'o'.

**8** Position the Scrubber Bar in the Timeline and use Overwrite to edit the shot into the Timeline.

Play back the edit and watch the result. If you have followed all of the above steps the Audio and Video will cut in the Timeline at separate places. If you marked a split for the end point Audio and Video will finish at separate places.

Achieving Split Edits in Final Cut Express is not the easiest of functions but it is definitely worth learning. Once mastered you can add finesse to a film and boost it into the professional realm. Any editor worth their salt will understand the process of performing Split Edits and know how and where to use them.

The Split Edit is also reflected visually in the Timeline. If you look at your audio and video tracks you will be able to see where the audio and video edits take place. If it is not clear you may have to use the Magnifying tool to increase the size of the Timeline and then each of the edits will be more obvious.

**Video 'in'**                                                       **Video 'out'**

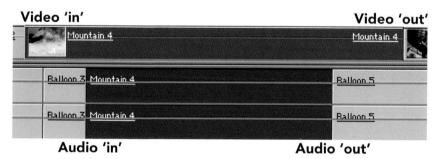

**Audio 'in'**                                                   **Audio 'out'**

If you find the above process difficult to follow there is another way to achieve Split Edits. This is done by Locking Tracks and editing directly in the Timeline with the Razorblade.

**1** Lock the tracks you do not wish to affect, for example lock your two audio tracks.

**2** Select the Razorblade tool from the Toolbar.

**3** Cut the video at the point where you want the Split Edit to occur.

**4** Choose the Pointer tool and highlight the piece of video you wish to remove. Press the Delete key.

5   Drag the end of the shot to fill the gap created when the razorbladed section was deleted.

You may find that it is not possible to drag the end of the shot to fill the gap. This could be because you are at the end of the media limit. All non-linear editing systems work by referencing to the original media stored on the hard disk. If all the media of a particular clip is already edited into the Timeline then it will be impossible to extract more frames.

Another way to achieve Split Edits is to switch off the yellow buttons in the Timeline and therefore restrict the flow of edited material. For example, if Audio 1 and Audio 2 are switched off only the video will flow through when performing an edit. Using the yellow buttons is like turning a tap on or off. If it's yellow it is Open – if it's gray it is Closed.

## Drag and Drop Editing

The editing we have done so far has involved opening a clip in the Viewer, marking an 'in' and 'out' point, and then editing this clip into the Timeline via the Canvas. It is possible to bypass this method altogether and edit clips directly from the Browser into the Timeline, or to edit from the Viewer to the Timeline without involving the Canvas in the equation.

1   Create a New Sequence in the Browser (press Apple N).

2   Double click the new Sequence to open it – you can rename the Sequence if you wish.

3   In the Browser click once on any of the clips – do not release the mouse button (open a Bin if all your clips are filed away).

4   With the mouse button still depressed drag the clip from the Browser directly into the Timeline. Release the mouse button and your clip will be edited into the Timeline at the position where you released your mouse button.

5   Do this again with a few more clips and you will see that it is possible to build a sequence simply by dragging clips from the Browser to the Timeline.

**6** Once you have several clips positioned in the Timeline repeat the procedure. However, this time drag the clip into the middle of the Timeline. Don't release your mouse button just yet!

When you have your clip positioned in the Timeline move the mouse button gently and notice that if you have the cursor pointing to the top third of the Video Track there is a horizontal arrow – if you point the cursor towards the bottom half of the screen there is a vertical arrow. A horizontal arrow represents Insert Edit whilst a vertical arrow represents Overwrite Edit.

Drag Clip to Top of
Video Track – Insert Edit

Drag Clip to Bottom of
Video Track – Overwrite Edit

The biggest disadvantage with using Drag and Drop editing is that you do not have control over marking your 'in' and 'out' points clip in the Viewer. However, this can be a very quick way to throw clips together into the Timeline. Furthermore, you can also do two other very useful tricks using Drag and Drop.

First, if you work in Picture Icon View you can arrange the icons in whichever order you choose (arrange them left to right in storyboard fashion) and then, by selecting an entire group of clips (by lassoing, or by using Alt and clicking to select multiple clips) you can then drag as many items as you wish into the Timeline. These clips will be positioned in the Timeline in order of the icon arrangement.

Once the clips have been dropped into the Timeline the Razorblade can be used to chop away unwanted sections. If you really wanted, you could edit in this way without ever opening a clip in the Viewer or dragging across to the Canvas (however, I would never rely on this exclusively as my method of editing).

# THE FOCAL EASY GUIDE TO FINAL CUT EXPRESS

**Highlight the Icons in the Browser and Drag these Directly into the Timeline**

**The Clips will then be Arranged in the Timeline in the Order they were Positioned in the Browser**

It is also worth noting that it is possible to drag clips from the Viewer to the Timeline, thus skipping out the step of editing across to the Canvas. The same rules apply as when dragging clips from the Browser to the Timeline. If you have a horizontal arrow this will represent an Insert Edit, while if you have a vertical arrow an Overwrite Edit will occur. You can also mark 'in' and 'out' points in the Viewer in the usual way. These 'in' and 'out' points will apply and therefore determine the beginning and end of the edit.

If clips are not open in the Viewer the 'in' and 'out' points will still apply. Thus should you drag a clip directly from the Browser into the Timeline, the duration of the edit and the start/end frame of the clip will be defined by the 'in' and 'out' points which have previously been set.

# Extending/Reducing Clips by Dragging

You probably feel like you have covered the editing process from every conceivable angle by now. There is one more useful way to work with clips in the Timeline, not yet discussed.

Clips can be made longer or shorter by grabbing hold of either end and dragging the length in either direction.

**1**   Choose a clip in the Timeline which you wish to extend or reduce. Position the Pointer and let it hover over the center of an edit point or the area where two clips meet. A symbol with two vertical lines and two arrows will appear.

**2**   If you wish to reduce the length of the clip drag the end of the clip into itself (using the symbol with two vertical lines). A display will appear to the right showing the overall clip duration and the trim adjustment in seconds and frames.

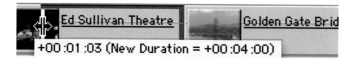

At the same time the Canvas will shuttle through the clip as you drag the end, giving you a visual reference to the adjustments being made. You therefore have a display in both the Timeline and Canvas giving you numeric and visual displays at the same time.

Providing you have Linked Selection switched on audio and video will move together, otherwise they will be independent.

3  In order to extend the length of a clip a gap must exist between the clip you wish to extend and the clip adjacent to it. To create a gap either insert a shot and then delete it and a gap will be created. Alternatively, use the Arrow tool to highlight and drag several clips further along the Timeline.

**Create a Gap**

**Extend the Length of the Clip into the Gap you have Created**

If you find it impossible to extend the length of a clip it most likely means there is no more material left to extract from the original clip. All clips have a media limit as defined by the amount of material originally captured to hard drive. Once this media limit is reached you can go no further.

For fine control when dragging, hold down the Alt/Option key. Another way to achieve fine control is to expand the Timeline by using either the Magnifier tool or the Slider Bar at the bottom of the Timeline.

## Single and Multi-Layered Effects

When it comes to building effects in Final Cut Express there are several things that need to be considered. First, there are the type of effects that are applied to individual clips or to a series of clips in the Timeline – these are single layer video effects. Then there are the types of effects that are termed multi-layered effects.

Look to the video track symbols at the front of the Timeline – next to the yellow symbols are V1 and V2. These stand for video track 1 and video track 2. Final Cut Express allows for up to 99 video tracks or layers to be created.

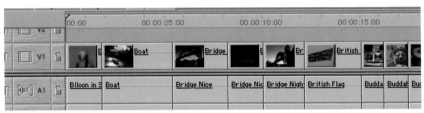

**A Single Layer of Video in the Timeline**

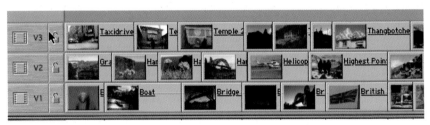

**Multi-Layers of Video in the Timeline**

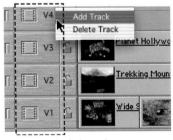

**Control Click in this Area**

Creating a video track – which means adding a layer – is as simple as control clicking between the little film icon and the locks in the Timeline and selecting the Add Track button. Tracks can be deleted using the same method and selecting the Delete Track command.

# EFFECTS

When working with a single layer of video, effects can be applied in the form of Transitions, Filters and Generators.

**Transitions** refer to an effect that is applied between two clips. Examples include dissolves, wipes and slides.

**Filters** can be applied to a clip or part of any clip. They are used to change the look of images by manipulating each of the video frames which make up the image. Examples include blurs, mattes, borders and changes to brightness, contrast and color.

**Generators** are neither Transitions nor Filters. These are used to create additional material which you need to work with. These include devices such as slug (black), text and matte colors. Think of Generators as being the electronic equivalent of spacer, head and tail leaders and countdowns of the film world.

When working with Transitions and Filters these are applied to a single layer of video and can be integrated into a project which is made up of no more than a single video track. Don't be deceived by this statement! Transitions and Filters can be applied to any layer within a project – the point is a project made up of no more than one layer can include Transitions, Filters and Generators. Multi-layered projects will also make use of Transitions, Filters and Generators and these can and likely will be applied to any of the layers in the Sequence.

An Example of Compositing Made Up of Two Video Layers

The instant you have more than one layer you are working in the realms of compositing. Layers are video components which are stacked on top of each other – examples include: picture in picture, titles and animated components.

Before we move onto Compositing it is important to know how to work with Transitions, Filters and Generators. You need to understand the difference between these. Just to refresh your memory: Transitions work between clips; Filters are applied to clips; and Generators work in a similar way to clips but are generated within Final Cut Express itself.

A Transition is Added Between Two Clips

A Filter is Added to a Clip

A Generator is Created Inside of Final Cut Express

## The Concept of Media Limit (Handles)

It is vital to understand when using Transitions that there must be available media for the Transition to work. This available media is referred to as 'handles'. The maximum length that a transition can be is equal to the available media or 'handles' which exist on the original clips as they were captured to your computer's hard drive.

Thus, should you wish to apply a 1 second dissolve between two shots then there must be at least 12 frames (PAL) or 15 frames (NTSC) of available media. The available media applies to the end (tail) of the outgoing clip, on one side of the transition, and the beginning or 'head' of the incoming clip. The idea is the same as checkerboarding two pieces of film in an optical printer or A/B rolling shots on separate machines in a linear tape suite.

If you do not have the available media then the length of a transition will be restricted to the media that is available. It is impossible to exceed these limits.

## Applying Transitions

**1**  Click once on the Effects tab in the Browser and this will reveal a list of Transitions, Filters and Generators. I suggest working in List Mode for this part of the operation.

**2**  Click once on the triangle to the left of the Video Transitions folder – this will reveal a list of the available Transitions. Choose the type of Transition you require and click on the triangle to the left. Now choose a Transition you want to work with.

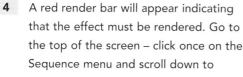

**3**  Drag the Transition onto the Edit Point in the Timeline where you want the effect to be applied.

**4**  A red render bar will appear indicating that the effect must be rendered. Go to the top of the screen – click once on the Sequence menu and scroll down to Render Selection. Release the mouse button and the render will begin.

Once rendered the Effect can now be played back and you can decide whether or not you are happy with it.

## Changing Transition Durations

**1**  Double click on the Transition in the Timeline – this will open a set of controls in the Viewer.

# THE FOCAL EASY GUIDE TO FINAL CUT EXPRESS

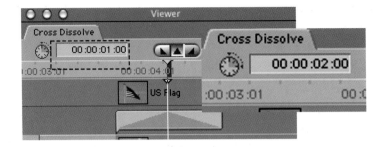

**2** At the top of the Viewer is the name of the Transition. Immediately below this is a numerical value representing the Transition duration. Click once with the mouse in this box and this will highlight the numeric value. You can now overtype this value. If you want a 2 second dissolve type 200 and press the return button; if you want a half second dissolve type 12 (PAL) or 15 (NTSC) and press return; for a 10 second dissolve type 1000 and press the return button.

If you prefer to view the Transition duration in frames hold down the Control key and click with the mouse on the duration display. This will open a contextual menu. Select View as Frames. Enter all durations in Frames from this point forward.

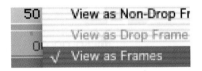

**3** The dissolve duration is now changed. You need to render the effect once again to play back the result. The above set of instructions applies regardless of which transition is chosen.

The Change in Transition Duration is Reflected in the Timeline

## Applying Filters

Filters are extremely powerful and creative tools to work with. It can take a good deal of experimentation to master them. It is definitely worth the effort,

EFFECTS

however. What can be produced on a desktop editing system with Final Cut Express would have cost a fortune at a facilities house only a few years ago. The price to be paid now is the amount of time the individual is prepared to spend building and rendering these effects.

**1**   Click on the triangle to the left of Video Filters in the effects area of the Browser. This will reveal a list of Filters stored in Bins. Open a Bin and choose a Filter you wish to experiment with.

**2**   Drag the Filter on to the clip where you want the filter to be applied. A red render bar will appear above the clip in the Timeline indicating it must be rendered.

**3**   To alter the parameters of the Filter double click the clip in the Timeline.

 Click on the Filter tab located at the top of the Viewer.

This will open a selection of controls that are used to alter the parameters of the Filter.

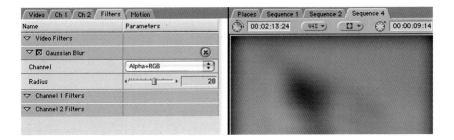

105

**4**  Make sure the Scrubber Bar in the Timeline is positioned on the clip to which the Filter has been applied. This is important as it ensures adjustments will be displayed visually in the Canvas.

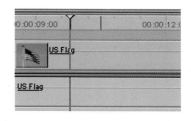

**5**  Experiment with the controls on the Filter tab in the Viewer and observe the result in the Canvas and onto your television screen (if you are set up with this configuration).

All Transitions and Filters can also be accessed from the Effects menu at the top of the screen. Simply scroll down to Video Transitions or Video Filters and select the effect you want. If you wish to use a Transition make sure the Scrubber Bar in the Timeline is positioned where the Transition is to be applied. If you choose a Filter, first highlight the clip in the Timeline and then choose the Filter from the Effects menu. Whether you choose to access the Effects from the Effects menu or from the Effects tab in the Browser comes down to personal preference. The result will be the same.

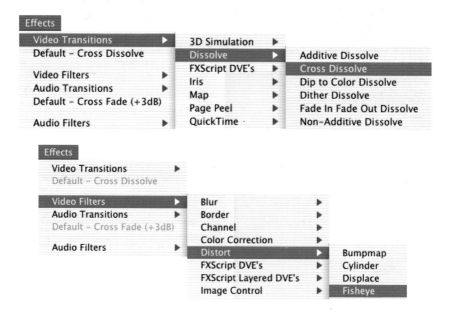

# Compositing

Compositing is where all the fancy stuff happens: flying titles, moving boxes, multi-layered dissolve sequences, transparent backgrounds – all the elements which 'dress up' a video production and make it more than cuts, dissolves, basic transitions and filters. As mentioned earlier, compositing encompasses everything that involves more than one layer of video.

Video filters, transitions and generators can all be applied to any video track, or several individual video tracks at a time. However, it is the 'stacking' of tracks or 'layering' of video that builds a composited sequence.

In on-line edit linear suites layering of video was done on a vision mixer in combination with a DVE (digital video effects). In the film world an optical printer was used. Pieces of film were sandwiched together (called bi-packing) and this was then exposed to several passes of light to achieve complex effects. Inside of Final Cut Express the tracks are layered in hierarchical order with the tracks closest to the top having priority over those below.

For those who have no idea what a DVE is – this refers to a stand-alone box used in television production for creating special effects. DVEs became popular in the late '70s and '80s and have been used all the way through to the present. DVEs have been seen as providing the video equivalent to the optical printer of the film world. They have traditionally been regarded as horrendously expensive, powerful devices. They are still used in live television production and on-line edit suites. However, as a result of programs such as Final Cut Express mere mortals can now achieve sophisticated effects with modest budgets. Previously a simple squeeze or flip (or flop as it is called in Final Cut Express) required one to step into an expensive post-production suite and often pay far more than the cost of Final Cut Express for a single effect!

# Methods of Creating Multiple Tracks

By default Final Cut Express opens with two video tracks in the Timeline. There are three possible ways to add more tracks:

1. Control click next to the V1 and V2 symbols (to the left of the locks). This will give you the option to either Add or Delete a track. If you click in the gray area above the last existing video track there will be a single option which is to add a track.

2. Drag a clip from the Browser or Viewer directly into the Timeline, into the gray area, above the top-most existing video track. Release your mouse button. This will then create a new video track where you drop the clip and two audio tracks below the last existing audio tracks.

3. Open the Sequence menu at the top of the screen and choose the Insert Tracks command from the menu.

Once you have created your Video Tracks items can then be edited to these tracks by dragging clips from the Browser or Viewer directly into the Timeline, or, you can direct the flow of video/audio using the yellow patching facility in combination with Insert and Overwrite editing.

Audio/Video will be Directed to V3 and A1/A2

EFFECTS

## The Motion Tab

The Motion tab is home to a great many of the tools used in the compositing process. In this window you will find the facility to resize images, rotate, reposition, crop, distort and adjust the opacity of clips. Effectively this provides you, the editor, with a fully fledged DVE facility. You can access up to 99 tracks which is equivalent to 99 independent layers of video.

To access the Motion tab, click the last tab to the right in the Viewer. Click the arrows to the left of each of the headings to access the controls.

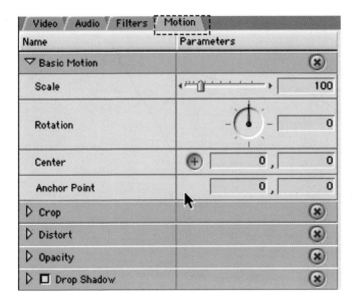

## Using the Motion Tab

To **scale** an image means to reduce or increase its size. The default size is 100% therefore numbers greater than 100 mean the image has been made larger or 'blown' up whereas numbers less than 100 mean the image has been reduced in size.

To alter the size of a clip using the Motion tab:

1   Double click a clip in the Timeline. This will open the clip into the viewer. Make sure the yellow Scrubber Bar is positioned on the clip in

109

the Timeline. This ensures you will see the result in the Canvas as you alter the parameters of the clip in the Motion tab.

**2** Click the Motion tab in the Viewer.

**3** Move the slider bar next to word **scale** either backwards or forwards. Alternatively type in a number. If you wish to reduce the clip in size by half type 50%. If you want to double its size type 200%.

Providing the clip is positioned on video 1 in the Timeline the result will be the image, reduced in size, over black. If you do not see this result you need to position the yellow Scrubber Bar over the clip in the Timeline prior to reducing the image in size.

Now repeat the process; however, this time reduce the size of an image on the V2 track over V1. This will create what is called a 'picture in picture'. The image on the V2 track will be reduced in size while that on V1 remains at 100%.

If you are working with the Real-Time effects switched on you will be able to see the result play in real-time in the Canvas when you play the clip. If you wish to watch the images play through the Firewire then render the result for playback.

To rotate images is easily achieved by selecting the **rotate** command underneath **scale**.

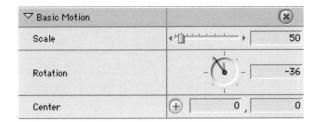

Simply turn the wheel and observe the result or type in a figure (remembering that a circle is made up of 360 degrees).

Again, to ensure the clip is active in the Viewer, first double click it in the Timeline – this will then make the clip active in the Viewer. Then click the Motion tab to see the controls, making sure the yellow Scrubber Bar is positioned on the clip you are working with. This will ensure the result is displayed in the Canvas.

To move a clip about within the Canvas click on the plus '+' symbol to the right of the word Center. A small '+' will then appear in the Canvas (you might have to look hard to see it – but it is there!). Click the '+' with your mouse and reposition the image within the Canvas window. Alternatively, type in co-ordinates into the boxes.

You may choose to rotate the image again or resize it. Other parameters can be altered. For example, check the **Drop Shadow** box towards the bottom of the Motion window.

Experimentation is the key. By adjusting parameters, typing numbers, adjusting opacity, altering angles – many different results can be achieved. No instruction manual can ever teach you everything that is possible. It is up to you to teach yourself through trying out the possibilities and learning by altering the controls.

## Image and Wireframe

An important and useful mode is what is called **Image and Wireframe**. Click the arrow in the center box of the Canvas and select Image and Wireframe.

Once you have selected Image and Wireframe you will notice a cross will appear from end to end across the active image. When working in this mode it is possible to slide the image around the frame by simply clicking on the center of the image with your cursor and then repositioning by dragging. This is a lot simpler than typing in center co-ordinates into the boxes or using the little '+' symbol as described on the previous page.

Click on Center of the Image and Reposition by Dragging

The image can be rotated by positioning your cursor on either edge and moving your mouse in opposite directions.

Your Cursor Becomes a Circular Arrow. You can then Rotate the Image

By dragging the corner of the image it can quickly be resized.

Position Your Cursor on the Edge of the Image and Drag to Resize

EFFECTS

Using Image and Wireframe provides a quick and effective way to reposition, resize and rotate images. The downside is you do not get the same control that you do by entering numerical values directly into the Motion Control window. Choose whichever option suits the task you need to perform.

## Titlesafe

Under the same menu as Image and Wireframe is another setting called Titlesafe. Both settings, Image and Wireframe and Titlesafe, can be switched on or off in either the Viewer or Canvas.

The **Titlesafe** setting is particularly important to have switched on when positioning images or working with titles. Domestic monitors do not display the full video image as it is recorded to tape. What is displayed on your computer monitor is known as 'underscan' while what you see on your television set is known as 'overscan'.

When Titlesafe is switched on it will be obvious – two sets of blue lines will be noticeable around the inside perimeter of the Viewer or Canvas. The Title Safe area is known as the Essential Message Area (EMA). In simple terms, to ensure that the images you are working with will be seen correctly on a television set, you need make sure they are positioned within the lines of the Title Safe area. The outer lines are regarded as safe on most televisions, whereas the inner lines are regarded as safe on virtually all television sets. So, when positioning images do not go outside of these lines unless you are comfortable that these images will likely be cropped to some extent on playback. It is important to remember that not all television sets are the same. Some will crop more than others.

## Working with Multi-Layers

It is relatively easy to create a picture in picture as already described. Using the same principles one can position several images on-screen at a time. The effect we will produce is to have a single image positioned as a background with three other images layered over it. Prior to systems such as Final Cut Express being available this could only be produced at facility companies and television stations using powerful DVEs and other expensive equipment. Now you can do it on your desktop quickly and easily.

EFFECTS

**1** Layer the images you want to work with over each other in the Timeline. Be aware that the order in which they are stacked will determine the order of priority of the layers. Thus the image on layer 4 will be at the front of the layers, while the image in layer 1 will serve as the background. Those on V2 and V3 will make up the middle layers.

**2** Reduce each of the images in size – I chose 35% – except for the background image – this stays at 100%. To achieve this you will need to individually click on each of the images to bring them into the viewer, and then resize accordingly. To achieve the same size for each image make sure the number in the Scale setting is the same.

**3** Make sure that Image and Wireframe is switched on and using your mouse position each of the images. This can be done manually or, if you prefer, use the '+' symbol to set X and Y co-ordinates in the Motion control window.

**4** If you find there is a black border on any of the edges as you place the images you need to use the **Crop** facility to remove this. Often a black edge is seen on the extreme outer perimeter of the video frame. This is never seen when the image is played to a television set. However, when an image is squeezed back this can become apparent. It needs to be cropped to get rid of it. Open the Crop controls and use the slider bars to crop the image; or enter a value to achieve the same result.

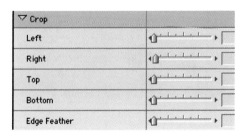

**5** Switch on the Drop Shadow in the Motion Control window. Controls for this setting are used to indicate the direction, distance, color, softness and opacity of the drop shadow.

115

If necessary, reduce the opacity of the bottom layer to make the images on the upper layers stand out. This is achieved by using the opacity slider in the Motion window or by switching on Clip Overlays at the base of the Timeline and then moving the bar which appears within the clip to the desired level. A window will count shifts in the level of opacity as you move the bar up and down.

The Slider Bar Affects the Opacity of Clips

Clip Overlays

Move the Black Bar Up and Down to Adjust Opacity

**Note:** if the opacity of the base layer of a sequence is adjusted the result will be a drop in luminance. Final Cut Express regards having no clip in the Timeline as being black. Therefore adjustments to the clip on the base layer will give the appearance of fading to black. Adjustments to further layers make these clips appear transparent.

Adjusting the Base Layer Causes a Drop in Luminance

Adjustments to Other Layers Cause Changes in Opacity

It is best to plan your effects before you begin to create them. By having a clear picture of what you are trying to achieve you have a much better chance of achieving something that works. By all means experiment. It is better, however, to experiment with vision and purpose rather than stumbling around in the dark hoping something acceptable will emerge.

## Keyframing Images

Keyframing allows you to set points which define a path which an effect will follow. An example of keyframing would be to move a box from one side of the screen to the other. These effects are built through use of layering and adjustments in Image and Wireframe mode. Neither transitions nor filters can be keyframed. However, adjustments to size, rotation and position can.

To keyframe an image decide first what you are trying to achieve. Let us work through the following example where we will have an image start on one side of the screen, rotate gently and increase in size until it comes to rest on the other side of the screen.

Two layers of video are involved in the above example. The first layer is a wideshot of Niagra falls with a boat in the distance; the second layer is a closer shot of the boat.

**1** To begin this effect you need two layers of video in the Timeline.

**2** Position the Scrubber Bar on the first frame of the Sequence and turn your attention to the Canvas. The video of layer 2 will be all you see for the moment. Using the Pointer double click the second Video Track in the Timeline to make it active in the Canvas.

**3** Select Image and Wireframe from the pull-down menu at the top of the Canvas.

4   Using your mouse grab the corner of the image and resize it.

Resize Image        Position Image

5   Click with the mouse in the center of the frame and drag the image to the left where you want the effect to begin.

6   Place the cursor on one of the sides of the image. The cursor will become circular with an arrow on one end. Move the mouse to turn the image.

Rotate Image

7   Press the Add Keyframe button located at the bottom of the Canvas.

8   Reposition the yellow Scrubber Bar in the Canvas (or Timeline) where you want the next keyframe point to be marked.

9   Click in the center of the image and drag it to the right. You will notice a line indicating the path the image will follow.

Reposition the Image. A Line will Indicate the Path the Effect will Follow

Resize and Rotate the Image

EFFECTS

10  Resize the image by dragging the corner.

11  Rotate the image by using the circular cursor on the side of the image.

12  Press the Add Keyframe button once the image is resized, positioned and rotated.

13  Once again reposition the yellow Scrubber Bar in the Canvas (or Timeline) where the next keyframe is to be marked.

14  Drag the image to its final position.

15  Resize and rotate the image.

16  Press the Add Keyframe button a final time.

It should be clear that a distinct set of processes is being followed each time a keyframe is added. The image needs to be sized, rotated and positioned – then the Add Keyframe button is pressed. The yellow Scrubber Bar is then positioned where you wish the next keyframe to take effect. Then the process is repeated – resize, rotate, position the image – mark keyframe. Move the yellow Scrubber Bar to the next position where a keyframe is to be added and so on… .

The duration of the effect is determined by the distance or separation of each keyframe as determined by the positioning of the yellow Scrubber Bar. The process is logical and straightforward. What is crucial is not to miss out any of the steps as this will interrupt the effect you are trying to create.

By dragging the Scrubber Bar through the effect in either the Canvas or the Timeline you can see the path the effect will follow. Render the result and check to see if you are happy with it. If not, repeat the process and try again.

Should you wish the box to start outside of frame you need to reduce the size of the Canvas using the drop-down menu at the top of the Canvas. This will resize the overview of the frame inside the Canvas giving you the flexibility of adding keyframes outside of the television viewing area.

By reducing the overall display to either 12% or 25% it is possible to have animated objects fly in and out of frame from the top, bottom or either side. By working with several layers of video one can animate several boxes at a time. The same can be achieved with text or imported Photoshop files.

## Multi-Layered Dissolves

Cuts and dissolves are the bread and butter of film-making. Effects have always been a luxury. However, with the availability of effects which don't cost money there has been a huge increase is what clients and producers demand. It seems as if everyone wants to add extra shine to what might otherwise be a mediocre film. While over-use of effects will fail to make a bad production good – tasteful use of effects can be pleasing to the eye and if nothing else provide visual interest. The term 'eye candy' has been used to describe many of the effects offered in desktop editing systems.

EFFECTS

A dissolve is nothing more than one shot fading into another. By placing a third layer over a dissolve it is possible to create what is known as the three-way dissolve. In the world of film production one had to plan dissolves, send the negative to the lab and hope the result was something close to what was envisaged. When using Final Cut Express you can experiment to your heart's content. You can even create four, five and six layer dissolves and beyond. The result can end up being a mash of unwatchable imagery – so be careful. Your reputation as an editor depends on it.

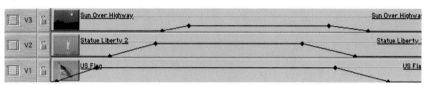

To create a three way dissolve is not difficult:

**1** Make sure you have three empty video tracks in the Timeline.

**2** Layer two clips on top of each other – the third layer will be added once you have set the opacity for the first two.

**3** Make sure Clip Overlays is switched on.

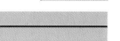

**4** Adjust the opacity of the video clips on V1 and V2 to achieve the result you are after. This is achieved by moving the black line inside the clip in the Timeline up and down. Providing the yellow Scrubber Bar is positioned within the bounds of the two clips the result will be displayed in the Canvas.

**5** Once you are happy with the result add the third layer of video to V3.

**6** Adjust the opacity until you can see all three images bleeding through. You may need to make further adjustments to 'get it right'.

**7** Render to check you are happy with the result.

**8** Select the Pen tool from the very bottom of the Toolbar.

9. Point your cursor at the black line of the base video clip. Your cursor will turn into a pen.

10. Click on the black line and a point will be added. This will be your first keyframe.

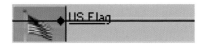

11. Add another keyframe further along. By allowing your cursor to hover over either keyframes you will see it turns to a cross.

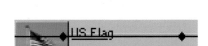

12. Drag the keyframe on the left to the base of the clip.

13. Perform the same function to the other clips, dragging keyframes to the base of the clip. What you are doing is plotting points so that the images will fade in and out at your discretion.

14. Once you feel you have the points correctly plotted, render the result and play it back. You may need to make further adjustments if the result is not as you wish it to be.

If you follow points 1–7 you will achieve the result of a three-way superimposition. This may be fine for your poses. If, however, you want the images to fade in and out at pre-defined points you need to use the Pen tool to plot keyframes. It is possible to have a single image play, then have another image dissolve over it to then be followed by a third image. I'm a big fan of three way dissolves. They can look great and produce subtle or high impact results depending on how they are treated. The interface within Final Cut Express is particularly suited to this type of work. The hassles required in a tape suite to achieve these effects were enough to persuade many editors never to try. As for the film world – now that would have been really difficult.

## Copy and Pasting Attributes

When you work hard building a set of effects it can be a great timesaver to be able to copy the settings from one clip to another. For example, you may wish to run a series of clips in slow motion at 35%. Rather than setting the speed for each clip individually it can be quicker to set this up for a single clip and then to copy the slow motion setting from one clip to all the others you wish to slow down. Details can be copied and pasted for many settings including: opacity, cropping, motion, drop shadows, filters and motion blur. These settings can be copied and pasted with just a few keystrokes.

**1** Control click in the Timeline on a clip from which you wish to copy the attributes. This will open a menu with many options. Select copy.

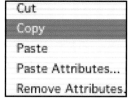

**2** Move your cursor to the clip which you wish to paste the attributes to. Control click on this clip and select Paste Attributes from the menu.

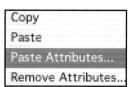

**3** Another menu will now open. Choose the options which you wish to apply to the clip. Do this by checking the boxes with a tick. Obviously the only attributes which you can apply are those which were already applied to the clip from which the attributes were copied.

# Titling

As an all-round editing package Final Cut Express covers virtually every area a film-maker will ever need to explore. No film would be complete without Titles. If nothing else a simple opening caption to identify a production is required. Final Cut does much more than this including Moving Titles, Transparent Shadows and Animated Text. Some of these operations can be quite complex; however, the basics are not difficult to master.

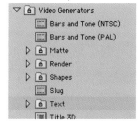

To access Titling in Final Cut Express go to the Effects tab and locate Video Generators.

1. Click the triangle to the left of the Video Generators to reveal a list of options.

2. Click the triangle next to Text – this will reveal six possible options for Text. For the moment choose the fifth option – Text.

3. Click on the Text Generator and drag it into the V2 track in the Timeline where you want the Text to be positioned.

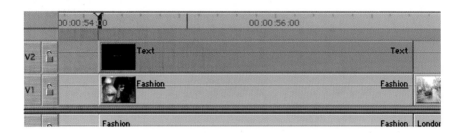

4. Double click the Text Generator on the second video track – this will load the Text into the Viewer. If you are working with Real-Time effects switched on a green line will sit above the line of text in the Timeline indicating the result will play in real-time. Otherwise, you will see a red line indicating that rendering is required.

5. In the Viewer click once on the second tab at the top – Controls.

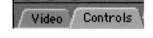

6. Make sure the Scrubber Bar in the Timeline is positioned on the shot that has the Text Generator above it. This ensures the result will be displayed in the Canvas as you work.

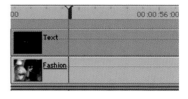

7. You should now see the words 'Sample Text' in the Canvas.

EFFECTS

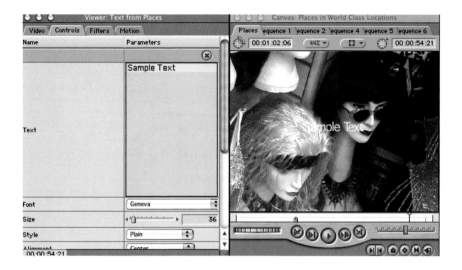

**8** Return to the Viewer – make sure you have clicked the Controls tab – click in the box to highlight the words 'Sample Text'. Overtype 'Sample Text' with whatever text you wish to enter.

**9** Click in the Timeline again and nudge the Scrubber Bar a few frames along using the arrow keys – the text will appear in the Canvas, over the shot where the Scrubber Bar is positioned.

125

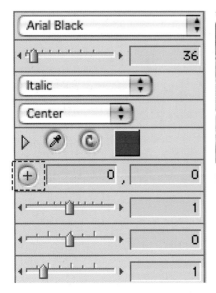

**10** You can modify the characteristics of the text by adjusting the details in the Controls area. You can change the font, the size, the color and tracking by altering each of the parameters.

**11** To reposition the text click on the '+' symbol in the Controls area labelled Origin. Click once on the '+' symbol and a small '+' will appear in the Canvas.

**12** Clicking in the Canvas window sets the '+' wherever you click with the mouse button. Release your mouse button at the location where you want the text to be repositioned. Alternatively, you can enter X and Y co-ordinates in the Origin area in the Controls window.

Another way to move the text around is switch on Image and Wireframe. Click with your mouse in the center of the active text window, and you can then freely position the text with the mouse.

There are other Text Generators available. These include Outline Text, Crawl, Typewriter and Scrolling Text. All of these are variations on the basic Text

Generator we have been working with. Experiment with these to see what they do and how they work. It is also possible to layer text, as with video tracks, so that several different layers of text can be built in different styles.

The easiest way to add a drop shadow is to go into the Motion window and switch drop shadow on. You need to make sure that the text generator is loaded into and active in the Viewer for this to work. You can then alter the direction, color, opacity and size of the drop shadow.

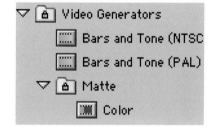

The quickest way to alter the overall opacity of the text is the same as with other clips in the Timeline. Simply switch on Clip Overlays and drop the black line inside the text generator in the Timeline to adjust the overall opacity.

To position text over a colored background select **Matte** from within Generators (found under the Effects tab in the Browser). You then need to overlay the Matte Color on top of the video track and then position the Text Generator of your choice over this. The

Matte Generator can be cropped in the Motion window and the transparency adjusted through either the opacity setting in the Motion window or by using Clip Overlays. Professional looking titles can then be achieved.

It is important to switch on the Title Safe generator and to position text within the bounds of the inner lines to ensure that no essential information is cropped during playback.

The only way to get good sound is to record it properly in the first place. Basic rules to achieve this include: using good microphones, getting the microphone as close to the subject's mouth as possible, listening to your sound through headphones as you record it, and setting your audio levels correctly. The likelihood is if it sounds good at the time of recording then it will sound good in the edit suite.

## Setting Correct Audio Levels

Audio levels are crucial to get right. If you set the level too high you will blow it, literally. When digital audio peaks too loud the sound will distort, break up and be unlistenable. Often called 'pumping', this will sound far worse in a digital environment than it would have in the analog world.

The basic rule with recording sound is don't let the audio meters push into the red. This applies for recording audio on location and working with sound in the edit suite. Many experts advise the correct level with DV audio to be no louder than −12 dB. I tend to allow my audio to peak between −12 dB and −6 dB and don't experience any problems.

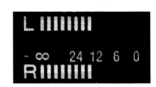

Do Not Allow Your Levels to Peak into the Red. Peak Between −12 dB and −6 dB

It is also recommended to keep audio levels lower rather than higher. Analog recordings tend to introduce hiss when the levels are raised – digital audio reacts differently. Obviously if hiss exists on the tape in the first place then this may sound louder when the level is increased. However, this is the result of an increase in the overall volume, rather than an increase in hiss level originating from the playback. By keeping the levels lower, rather than louder in a digital environment, you are less likely to experience problems.

## Getting the Most Out of Your Audio

Once your audio is recorded you are more or less stuck with it. While it is possible to improve the quality through use of filters and other means, in general the quality is determined by what is recorded on tape in the first place. However, within Final Cut Express there are several features which enable you to put together a good sound mix. The sound mix refers to the way all of the elements blend together to create the overall soundtrack.

To produce an integrated soundtrack which is both seamless and effective one needs to be able to adjust audio levels and to program smooth fades and mix several tracks of audio together. Final Cut Express allows the editor to adjust and mix audio levels in real-time (up to eight tracks) and it does a very good job of this.

To mix audio effectively I strongly suggest that you work with what is known as Stereo Pairs. Once captured, clips can be converted into Stereo Pairs.

## Converting Clips into Stereo Pairs

Each clip that you capture is made up of two tracks – a left track and a right track. When mixing these tracks, unless you are working on a complex soundmix with defined stereo separation, it is useful to marry these audio tracks together so that any adjustments to audio levels will apply to both tracks. Otherwise, when you adjust the audio levels you will need to make sure that each track is adjusted by exactly the same amount – a difficult and time-consuming process.

You can tell if your clips are Stereo Pairs by looking at the audio tracks in the Timeline. A Stereo Pair is defined by two sets of triangles facing each other. If these triangles are present then you are working with Stereo Pairs – if there are no triangles present you need to convert your tracks into Stereo Pairs.

| **1** | Select the horizontal Arrow tool in the Toolbar and highlight the entire contents of the Timeline. |

| 2 | Choose the pull-down menu at the top of the screen titled Modify. Scroll to Stereo Pair and release your mouse button. This toggles to Stereo Pair. A tick means Stereo Pair is selected. By selecting Stereo Pair you are instructing Final Cut Express to convert whichever clips you have highlighted. | 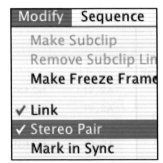 |
|---|---|---|

| 3 | All your clips should now be converted into Stereo Pairs. You can confirm this by checking that two sets of triangles facing each other are present in each of the audio tracks. Providing these triangles are present then your audio has been converted into Stereo Pairs. | 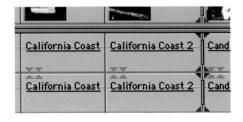 |
|---|---|---|

It is important to listen to the sound of the clips once they are converted. Then compare the sound to the clips before they were converted to Stereo Pairs. I have experienced on some occasions times when clips sound better as non-stereo pairs. To compare simply convert a clip to a Stereo Pair, listen to it, then press Apple Z to undo the conversion and listen to the result.

## Adjusting Audio Levels

| 1 | Click on the Clip Overlays symbol at the bottom left of the Timeline. |  |
|---|---|---|

2. A pink line appears in each of the audio tracks and a black line in the Video track. You will be familiar with Clip Overlays from the effects section.

3. Point your cursor at the pink lines. When you get close to these lines your cursor turns into two short horizontal lines with vertical arrows on either side.

4. Click your mouse button and you can now move the pink lines up or down. If you move the lines up the volume for the clip will increase, if you move the lines down the volume will decrease. Providing you converted the clips into Stereo Pairs the lines will move together as you make adjustments. Using this method it is possible to balance any differences in audio levels between clips to achieve a smooth and natural sound mix.

As you make adjustments keep an eye on the audio levels on the VU meters within Final Cut Express or, if you are working with a deck or a camera, on the meters on the external device.

Whilst it is most useful to be able to adjust audio levels in this way the requirements of most productions go far beyond being able to increase and decrease the sound levels. One needs to be able to include smooth fades and cross fades to build an audio mix where each of the elements blend together in a perfect mix of sound effects, narration and music.

## Adding Sound Fades

1. Make sure Clip Overlays is switched on and make sure all the clips in the Timeline have been converted to Stereo Pairs.

2. Select the Pen tool at the very bottom of the Toolbar.

3. Point your cursor at the pink lines and the cursor will turn into the Pen tool. Choose a point on the pink lines where you wish for the sound fade to begin and click with your mouse. A pink dot will be marked. This mark is a keyframe and represents the beginning of the fade (the keyframe will apply to both audio tracks in the Stereo Pair).

4   If you want to reposition the keyframe, point the cursor at the keyframe mark and the cursor now becomes a small cross (+). Click once, with the small cross on the keyframe. You can now move the keyframe by dragging.

If you wish to delete the keyframe hold down the Alt key and your Pen tool will now have a minus symbol next to it. If you click on the keyframe with the minus symbol the keyframe will be deleted. The Pen tool with a minus symbol can also be selected from the Tool Palette by extending the Pen tools and choosing the second option.

5   Add a second keyframe further along in the clip. You should now have two keyframes marked. To create an audio fade at least two keyframes are required.

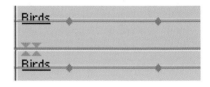

6   Hold the Pen tool over the first keyframe so that the tool becomes a small cross (+). Use the little cross to drag the first keyframe down to the base of the clip. You should now have a curved line which starts at the bottom of the clip and rises to a point at the top of the second

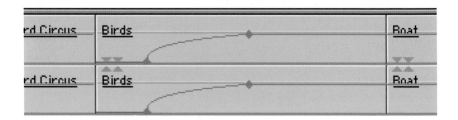

keyframe. Play back the clip and your sound will rise from silence at the first mark to a defined volume at the second mark.

For fine adjustments hold down the shift key while dragging the keyframes up or down.

It is useful when working with audio to increase the size of the Timeline vertically, thus giving a better view of any adjustments you make. This is achieved by clicking on the small boxes located on the bottom left-hand side of the Timeline. Choose whichever size you feel most comfortable with.

It can also be useful to increase the overall horizontal size of the Timeline. Use the Magnifying tool to achieve this or pull on the ribbed ends of the slider bar at the base of the Timeline. This allows for fine adjustments to be made to the audio over time.

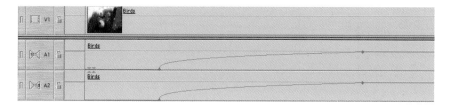

Increase the Spread of the Timeline for Fine Control

## Adding Audio Cross Fades

Audio Cross Fades are used for creating seamless blends between Audio Transitions. All sorts or unwanted sounds can be easily eliminated.

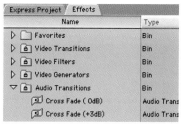

1  Select the Effects tab located top right of the Browser.

2  Scroll down to Audio Transitions – click on the triangle to the left to reveal the contents.

THE FOCAL EASY GUIDE TO FINAL CUT EXPRESS

3   Click on Cross Fade (0 dB) and drag this to the cut point between the two audio tracks where you want the Cross Fade to occur.

4   Playback the section with the Cross Fade and decide whether you are happy with the result. If you wish to change the duration double click on the Cross Fade which is positioned at the cut point between the two clips. This will open a dialog box, which will allow you to enter a new duration. Enter the duration and click OK.

The result should be a nice smooth Cross Fade where one section of audio blends into another.

Audio Cross Fades can also be added by positioning the yellow Scrubber Bar at an edit point and choosing the Effects menu at the top of the screen. Choose one of the Audio Transitions.

A third way to add Audio Cross Fades is to Control click at an edit point between two clips. The option will be given to add a cross dissolve if you click in the video track or to add a cross fade if you click in either one of the audio tracks.

## Adding Audio Tracks

Audio tracks can be added or deleted in the same way as video tracks. Simply hold down the Control key and click in the gray area of the Timeline next to the audio symbols. A menu will open giving you the option to add or delete a track. Remember to add two tracks for each section of Stereo Audio required.

Just as video tracks can be dragged directly into the Timeline so can audio tracks. If you wish to add a CD track, for example, which you have already imported into the Browser then drag the track directly from the Browser into the Timeline. If the item is dragged to an already existing audio track you will get the result of an Insert Edit if your cursor points to the top third of the track and an Overwrite Edit if your cursor points to the bottom half.

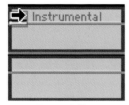     

        Insert Edit                      Overwrite Edit

Audio tracks can be added to the Timeline by dragging a CD track directly from the Browser to the gray area in the Timeline below the last existing audio track.

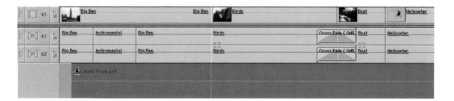

The CD track will then be added and two new tracks created to accommodate it.

This is the quickest way to create new tracks and at the same time get a new piece of audio into the Timeline. If you wish you can then delete the audio and these tracks will remain free to be used however you wish.

## Mixdown Audio

If you work with more audio tracks than your computer can comfortably play then you may choose to invoke the Mixdown Audio command. This will mix all your audio tracks into a single file. Everything will appear the same in the Timeline. However, Final Cut

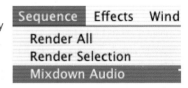

Express will reference to a single file thus getting around the problem of not being able to handle such a complicated sound mix. The downside of using the Mixdown Audio command is that each time you change the edits in the Timeline you will need to do another Mixdown Audio. Unless you are working with many tracks of audio, or use a lot of audio filters, keyframes or effects, then you should not run into problems.

## Playing it Safe

You've edited your movie, all the effects and titles have been added and you have perfected your audio mix so that everything is just right. Now it is time to output your movie.

The simplest form of output is to roll record on your camera or deck and then play the movie directly out of Final Cut Express through the Firewire and straight to tape. In fact this is something you should do regularly throughout the editing process to ensure you have a backup of your film should a technical catastrophe or any other sort of problem render your project useless.

I always back up my material to digital tape in the form of several different versions: with and without graphics and effects; and other individual passes to tape with separate audio tracks, both mixed and unmixed. One pass may include voiceover, another music, and another sound effects. By doing backups in this way all of the raw elements of the film are preserved so that in the event of a corrupt file, human error, or power failure at a critical moment, the most you will lose is the work you have done since your last backup. It takes discipline to backup your work regularly; however, you will pay a serious price and learn the hard way if you do not.

## Print to Video

Print to Video is a function Apple built into Final Cut Express to give a professional look to a finished film. Here the project can be named, color bars can be inserted at the head and black can be added at the end of the production. It is also possible to loop the film so that several copies can be recorded onto a single digital tape, or one can choose to record sections of the movie by defining 'in' and 'out' points and only printing these sections to tape.

Print to Video also serves another purpose. If any material is unrendered then the computer will render this material prior to invoking the Print to Video command and if your audio mix is complex then a mixdown will take place. This ensures all the components of your production should play without problem. Sometimes dropped frames may be encountered during normal playback and the problem will be cured when the Print to Video instruction is given.

OUTPUT

Selecting Print to Video is often the final stage of the editing process. Think of it as getting the release print off to the lab once the hard work in the cutting room has been done.

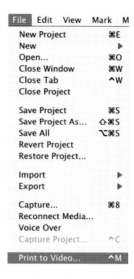

**1** Select the File menu at the top left of Final Cut Express and scroll down to Print to Video.

**2** Release the mouse button and a box will appear giving you many options to choose from. Check those boxes which apply to your specific needs.

Add color bars if you wish and enter a duration for the amount of bars you wish to record. Instruct Final Cut Express if you want black to be recorded after the bars. Enter a title for your production in the text column and select Print Entire

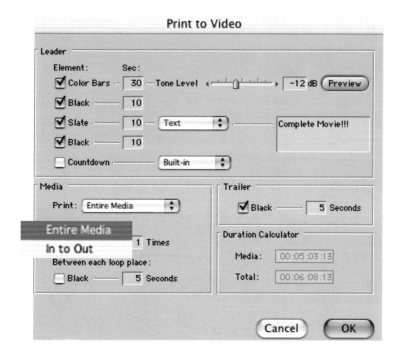

141

Media or Print In to Out depending upon your requirements. If you choose the Print In to Out option you need to mark 'in' and 'out' points in the Timeline.

3   Once you are satisfied that you have correctly selected the options you require click the OK box. The computer will pause as all the elements are gathered together.

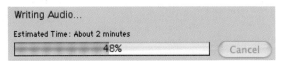

Now sit back and enjoy your movie. It is worth keeping a close eye on the output of Final Cut Express. Print to Video is the last stage in the editing process (unless you are using it to label sections of a work in progress) and therefore it is wise to make sure everything is exactly right as the signal is recorded onto tape. Once you have at least one master safely dubbed you can relax knowing that the vital information has made it from the original camera masters, onto your hard drives, through the editing process, and finally back onto digital tape. Don't forget to do a second and maybe even a third backup. This may just help you to sleep better at night.

## Other Forms of Distribution

Final Cut Express is designed as a DV editor. However, just because it is designed to work with DV does not mean that your final output must end up on DV tape. There are other formats of distribution in this modern world that need to be considered; for example, CD-ROM, internet delivery and DVD.

For most of these purposes one would acquire and edit the material on DV or DVCam and output the master edit to DV tape before compressing the final product to another format.

QuickTime is a cross-platform (Mac and PC) technology designed and developed by Apple and this is often the format of choice for both internet delivery and CD-ROM production. There are other options and other competing standards – I would suggest using QuickTime for the simple reasons that (1) you are working on a Mac and QuickTime is the video architecture built into the Mac OS operating system, and (2) QuickTime is a high quality, cross-platform standard particularly suited to video production. However, before choosing the exact form of compression, you need information as to how and where the product will be used.

When outputting to DVD, MPEG-2 is the compression format which must be used. While Final Cut Express does not provide an in-built facility to output to MPEG-2 this is catered for with other Apple programs such as iDVD and DVD Studio Pro.

When compressing video for the web, CD-ROM or DVD decisions must be made about what level of compression will be used. There are always compromises with regards to the level of compression which needs to be considered. When preparing material for internet delivery the goal is to reduce the overall file size to a minimum while maintaining the best possible quality. Most consumers use a 56k modem and therefore sending gigantic video files down the line is not an option. Even still images are compressed heavily for web use and video is made up of thousands and millions of still images. Factors which need to be considered include what information is in the video frame. A static camera shot like a talking head, for example, will compress much better for web delivery than images with loads of camera movement.

The purpose of this section is not to give detailed instructions of how to achieve these forms of output, what levels of compression to use, and exactly how to go about this – but rather to give you an overview of the areas you need to explore to achieve results. There are entire books written about producing DVDs, preparing material for internet delivery and CD-ROM production – therefore it is impossible to do justice to this subject in a few pages. Rather I will explain how the compression process works when outputting from Final Cut Express. To achieve final results will require further work on your behalf.

Once you have edited your movie there are two areas you need to look to – both are found under the File menu. Scroll to Export and you will see there are two choices: Final Cut Movie and QuickTime.

Choose the option Final Cut Movie. A window will open with further options. You can choose to export video, audio or both. It is particularly important to pay attention to the Make Movie Self-Contained command. If this box is ticked

it means that every frame in the Sequence which is being exported will be recompressed, a very time-consuming process, and this results in an entirely new file of your movie being created which will be large in size. The alternative is to uncheck the Make Movie Self-Contained box. The file which is created is then a reference movie. This file will be quite small, quick to create, and will reference back to the original files which your movie has been created from. It is therefore desirable to use the Export as Final Cut Movie option and to create a movie which is a reference movie (uncheck the tick in the Make Movie Self-Contained box). This file can then be used in applications such as iDVD and DVD Studio Pro as a source file to tell the program that this is the movie you wish to work with.

To export a Final Cut Movie you need to highlight a Sequence in the Timeline; choose the Export as Final Cut Movie option; check through the possible options such as whether you are working with video, audio or both and then save the Sequence to the desktop.

Exporting a file as a QuickTime movie requires more knowledge on your part. There are a far greater number of choices and possible options so you really need to know what you are trying to achieve before you start. It is worth opening up the Export as a QuickTime facility to explore what the options are.

As already mentioned, QuickTime is part of the Mac OS operating system – QuickTime has been described as the backbone upon which all video on the Mac rests – it is what makes video possible. The technology is cross-platform and is suitable for web and CD-ROM delivery. You can rest assured if you create a file using QuickTime it will work in all Macs and also in PCs, providing QuickTime is installed on the PC. However, there are options to consider when compressing a file from DV to QuickTime.

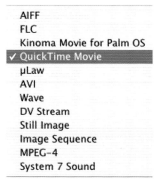

Choose the Default Settings box at the bottom of the window and you will see you have many choices. The most important advice I can give here – and I know I am repeating myself – is to know what you are trying to achieve. There are many options to consider and the choice of options will affect file size, image quality, audio quality, the frame rate and whether playback of images will be smooth or jerky.

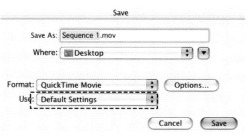

You need to experiment. Look through the options, make a choice, export a movie and see the results. You can set different levels of compression, different sample rates for sound, decide whether the movie is to be prepared for Internet Streaming or not. There is no easy option.

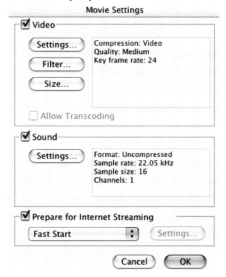

145

Furthermore, there are other possibilities to consider beyond simply choosing QuickTime.

Audio, for example, can be exported as an Aiff file. When exporting as an Aiff file consider: is the sound mono or stereo? what is the sample rate?

AIFF
FLC
Kinoma Movie for Palm OS
✓ QuickTime Movie
μLaw
AVI
Wave
DV Stream
Still Image
Image Sequence
MPEG-4
System 7 Sound

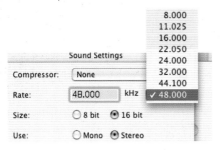

Images can be exported as Still Images. Choose PNG, Tiff, Pict or Photoshop to name just a few of the options.

If this all sounds confusing it is because we are dealing with a confusing subject – video and audio compression is a subject which is constantly evolving. However, having access to the tools gives you access to the knowledge you need to work in this area. The beauty of working on the computer is you always get a second shot at it, and a third, fourth and hundred and thirty first shot if you need it. A little bit of knowledge will take you a long way and your knowledge will accumulate as you go. Remember – knowledge is power! The power of distribution is at your fingertips through the process of compression. Press the buttons, wait for the encoding, and see the result.

## Final Cut Pro vs Final Cut Express

The big thing that Final Cut Express lacks, which Final Cut Pro has, is that Final Cut Express is only designed to work with DV – or Digital Video. Final Cut Pro has been built so that it can edit any video or film format on the planet.

There are many similarities between Final Cut Pro and Final Cut Express. However, with the release of Final Cut Pro 4 it is no longer fair to call this an application or piece of editing software – it is now a suite of applications. Here's what you get.

**Final Cut Pro** – The editor is similar to Final Cut Express; however, it is more colorful with a bunch of workflow improvements. Features include Batch Capture, Real-Time effects that play through the Firewire, the ability to work with low resolution footage through a function known as Offline RT, advanced Color Correction and professional media management tools. These are just a few of the features which Final Cut Pro 4 offers over Final Cut Express.

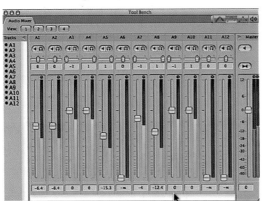

There is an integrated audio mixer with faders which will plot keyframes as changes in audio levels are made. The keyboard is now fully 'mappable' which means an editor migrating from another system can easily program the keyboard to emulate the setup they are used to; and there is also a feature known as Time Remapping which allows clips to be 'ramped' into slow motion or fast motion in variable amounts.

**Cinema Tools** – This used to be sold as a separate application which retailed for the same price as Final Cut Pro. It is now integrated as part of Final Cut Pro 4. For those shooting and outputting to film, Cinema Tools is used to conform the original camera negative to the edit produced within Final Cut Pro. What this means to the world of film editing is that the cost of post-production is now a whole lot more affordable than it used to be. For those planning to shoot and edit a production which originates on film this feature will be worth the cost of Final Cut Pro 4 alone – for the rest of us it will unlikely ever get used.

**Livetype** – Text has been criticized as being the flaw within Final Cut Pro in the past. I would describe the text facilities offered with Final Cut Pro 4 as the best I have ever seen. Using Livetype it is possible to animate text in a myriad of ways using pre-programmed templates in combination with textures and backgrounds. The problem with templates is it usually means being locked into a closed set of parameters. Not the case with Livetype. The templates can be easily modified and these results look great. You can even do flying text with flames shooting out of it.

**Soundtrack** – What Apple have provided with Soundtrack is a royalty-free music library from which the editor can access all the components needed to make up a musical score: loops, riffs, backing tracks, drums, changes in tempo, changes in key – it's all there. Edited music can be made to fit any length – which is great for 15 or 30 second spots – and the music can be layed out in such a way to peak and fade according to the action and pace of the edited sequence. The interface has been clearly modeled on the look of Final Cut Pro with a familiar looking Timeline and a window which plays video in sync with the building blocks which make up your musical score.

**Compressor** – This is an extremely high quality software encoder which provides for batch export of files to MPEG-2 for DVD authoring or MPEG-4 for internet delivery. All QuickTime formats are also supported. Compressing video might not be the most exciting task of editing. However, in the professional world it is vital to be able to do this efficiently and at very high quality and due to variable bit rate encoding (VBR) Compressor does just that.

It is also worth noting that Final Cut Pro 4 will only run on a G4 processor machine whereas Final Cut Express will run on G3 and G4 processors.

The fact is that most people who initially bought Final Cut Pro did so with the intention of using it to work with DV. Final Cut Pro has always worked with formats outside of the DV spectrum – now with the release of Final Cut Pro 4 the integration is complete.

The key feature of Final Cut Pro 4, which for me stands above the many other new and extremely desirable features, is that through Firewire 800, or Firewire 2, one will now be able to plug a DVC Pro 50 camera directly into the Mac. Effectively, uncompressed video will now work straight into the Mac. This is a huge step forward for film-makers wishing to work with formats of higher quality than standard DV. Furthermore, for those wishing to use other uncompressed formats such a Digital Betacam, third party devices are now hitting the market which will allow uncompressed video to be sucked straight into the Mac through a standard Firewire 400 port.

It is my prediction that Final Cut Pro will become the dominant editing standard throughout the world. The facilities offered are unparalleled by any other editing system at a price that is within reach of anyone earning an average wage. So where does this leave Final Cut Express users? I would say in a very good position. Final Cut Express offers the perfect means to become acquainted with the art of film-making, at an extremely affordable price. If you outgrow this software, or find you need the features which Final Cut Pro offers, then you will have no problems migrating to the full version of the program. The interface is so similar that within minutes you will be editing with confidence (although it will take longer to learn all of the new features). The fact is, for many DV editors Final Cut Express will serve their needs. If it doesn't, step up to the full version and learn to fly a full-sized Jumbo rather than a 737.

# Epilog

Since home computers appeared in the market-place Apple computers have consistently been at the heart of creative environments the world over. It is therefore no surprise that Apple are now at the center of what has been dubbed the digital video revolution.

In just a few short years Apple have created an editing system which has infiltrated the market at all levels. This is quite an achievement for a platform which many predicted would be extinct by the turn of the last century.

So the big question is what will be next? What lies in store for us with the Mac platform and video editing.

Undoubtedly there will be faster chips, bigger hard drives, great looking screens and innovative designs. Perhaps we will see quad processor machines, high definition video written straight to DVD and high-speed internet connections which make the current systems seem like delivering mail by horse and coach.

Apple have always been a secretive company, only announcing product releases and new strategies at moments of their choosing.

Many people ask me when will the G6s, '7s, '8s arrive, when will we see faster chips, what is the next big thing going to be… .

The moment is now. We have great tools to work with; superb tools; incredibly sophisticated tools. Final Cut is as good as it gets.

Seize the moment.

Tell your stories.

Now.

# Index

16 Bit audio, 26

Active tracks, 67
Adding tracks, 69, 100, 108, 136–8
Arrow tool, 70, 75–6
Aspect ratio, 9, 28
Attribute copying, 122–3
Audio:
  Cross Fades, 135–6
  effects, 129–38
  fades, 133–6
  Final Cut Pro, 148, 149
  levels, 16–17, 130–3
  Meters, 16–17
  Mixdown, 138
  mixing process, 3, 4
  mute, 22
  output formats, 146
  peak levels, 17, 130
  sample rates, 26–7, 40–1
  sound quality, 130–1
  Stereo Pairs, 131–2
  tracks, 67–9, 136–8
  Transitions, 135–6
Autosave Vault, 23

Backups, 23–4, 140
Batch Capture, 36–7
Bins, 48–50
Blurs, 101
Borders, 101
Browser, 16–17, 45–7

Cameras, 5, 6–7
  computer control, 31–2
  playing video, 44–5
Canvas, 16–17, 60
Capture, 3, 4, 31–9
Capture Clip, 31, 34–5
Capture Now, 31, 35–6
CDs:
  CD-ROM output, 142–3
  copying tracks, 38–41, 137
  sample rate, 27, 40–1

Cine-film, 2–3
Cinema tools, 148–9
Clip button, 35
Clip Overlays:
  audio, 132–3
  opacity, 116, 127
  symbol, 22
Clips:
  see also Editing
  bins, 48–50
  capture, 34–6
  copying attributes, 122–3
  deleting, 50–1
  duration, 33
  finding, 52–4
  lengthening/shortening, 97–8
  names, 35, 36
  playing speed, 83–4
  size, 22, 109–10, 112
  Stereo Pairs, 130–2
  Subclips, 37, 87–8
  Thumbnails, 45–6
  viewing, 44–5
Close Gap, 74
Color bars, 140–1
Compositing, 3, 4, 101–2, 107–27
Compression, 142, 145, 149
Computer controlled
  DV Deck, 31–2
Connectors, 7–8, 45
Control track, 37–8
Conversion:
  audio sample rates, 40–1
  video formats, 31, 150
Copying:
  see also Capture
  attributes, 122–3
  CD tracks, 38–41, 137
  clips, 76, 77
Crop, 115
Cross dissolve, 136
Cross Fades, 135–6

153

# INDEX

Custom Layout, 19–20
Cut, 77

Decks, 5, 7, 31–2, 44–5
Deleting:
  clips, 50–1, 74
  tracks, 69
Digital Video (DV):
  decks, 5, 7, 31–2, 44–5
  DV-NTSC, 8, 88
  DV-PAL, 8, 88
  effects (DVEs), 107
  Start/Stop Detection, 47–8
Dissolves, 101, 120–2
Distribution, 4, 139–46
Drag and Drop editing, 94–6
Dragging:
  clip extension/reduction, 97–8
  Timeline expansion/contraction, 62
Drop Shadow, 111, 115, 127
Dropped frames, 6, 140
Duration:
  clips, 33
  transitions, 103–4
DV *see* Digital Video
DVD outputs, 143
DVEs *see* Digital Video (DV), effects

Easy Setup, 27–8
Edit menu, 77
Editing, 3, 4
  basic process, 57–65
  clip extension/reduction, 97–8
  copy and paste, 76, 77
  Drag and Drop, 94–6
  individual tracks, 67–9
  Linked Selection, 71–3, 78–9
  locked tracks, 68–9, 73, 79, 93
  Match Frame, 89–90
  original process, 2
  playing speed, 83–4
  snapping, 77, 78
  split, 91–4
  sync, 80–1
  undo/redo, 71
Effects:
  audio, 129–38
  video, 99–127

Eisenstein, Sergei, 2
Essential Message Area, 113
Extending clips, 97–8

Fades, 133–6
Fast motion, 83–4
File types, 142–3, 146
Film structure, 20–1
Filters, 101, 102, 104–6
Final Cut Movie, 143–4
Final Cut Pro, 148–50
Finding:
  clips, 52–4
  shots, 90–1
Firewire, 7–8, 27–8, 44–5
Fit to Fill, 65
Fit to Window, 22
Formats:
  input, 31, 150
  output, 142–3, 146
  video, 8–9, 13, 27, 30–1, 150
Frames:
  dropped, 6, 140
  freeze frame, 88–9
  Keyframes, 117–20, 133–4
  Match Frame Editing, 89–90
  number per second, 88
  Poster Frames, 45–6
  Wireframe, 112–13
Freeze frame, 88–9

Gaps, 74–5
Generators, 101, 102, 123–7

Handles, 102
Hard drive, 6
Hardware requirements, 5–8, 149

Image manipulation *see* Video
Image and Wireframe, 112–13
'In' point, 33, 35
  editing, 57–8, 64
  modifying, 66
Input formats, 31, 150
Insert Edit, 56–64, 73–4, 76
Installation, 12–14

154

Interface layout, 16–20
Internet formats, 143

Join Through Edit, 79

Keyboard commands:
  DV Deck, 31–2
  editing, 64
  'in' and 'out' points, 66
  scrubbing, 57
  undo/redo, 71
Keyframes, 117–20, 133–4

Laboratory, 3
Letterbox format, 9
Linked Selection, 71–3, 78–9
List mode, 51–2
LiveType, 149
Locked tracks, 68–9, 73
  Razorblade tool, 79
  Split Edits, 93
Luminance, 116

Magnifier tool, 70, 79–80
Make Movie Self-Contained, 143–4
Make Subclip, 37, 87–8
Match Frame Editing, 89–91
Matte, 101, 127
Media limit, 102
Memory requirement, 5
Mixdown Audio, 138
Mixing, 3, 4, 130–8
Monitoring buttons, 22
Motion tab, 109–11
Moving images, 117–20
MPEG-2, 143
Multi-layered effects, 114–22
Multiple:
  images, 114–16
  sequences, 81–2
  tracks, 108
Music, 38–41, 149
Mute button, 22

Names:
  bins, 49
  clips, 35, 36
  project, 140
  sequences, 82

Now button, 36
NTSC, 8, 88

Opacity, 116, 127
Operating system, 5
'Out' point, 33, 35
  editing, 58, 64
  modifying, 66
Output, 4, 139–46
Overlays, 22
Overwrite Edit, 56–64, 73, 76

PAL, 8, 88
Paste, 76, 77, 122–3
Paste Insert, 77
Peak audio levels, 17, 130
Pen tool, 70
Picture in picture, 110–20
Pointer tool, 70, 73
Poster Frames, 45–6
Primary Scratch Disk, 14
Print to Video, 140–2
Processes, 3–4
Processor requirements, 5, 149
Pumping, 130

QuickTime, 142–3, 144–6

Razorblade tool, 70, 78–9
Real-Time effects, 86–7
Record monitor, 16–17
Redo, 71
Reducing clips, 97–8
Reference movie, 144
Release prints, 3–4
Rendering, 84–6
Replace Editing, 65
Reverse play, 84
Rotation, 111–13

Sample rates:
  CDs, 27, 40–1
  DV audio, 26
Saving projects, 23–4, 140
Scale, 109–10
Scene Detection, 47
Scratch disks, 28–30
Screen layout, 16–20
Scrubber Bar, 44, 58

# INDEX

Searching:
  clips, 52–4
  shots, 90–1
Selection:
  items, 70
  linked, 71–3, 78–9
  multiple items, 75–6
  tracks, 70
Self-Contained, 143–4
Sequences, 57–60, 81–2
Setup, 12–14, 27–8
Shadows, 111, 115, 127
Size:
  clips, 22
  image, 109–10, 112
  Timeline, 62, 79–80, 135
Skipping, 77
Slider tool, 62
Slides, 101
Slow motion, 83–4
Slug, 101
Snapping, 77, 78
Software installation, 12–14
Sound:
  *see also* Audio
  mixing, 3, 4, 130–8
  quality, 130–1
Soundtrack, 149
Source monitor, 16–17
Speed, 83–4
Split Edits, 91–4
Standards, television production, 8–9
Start/Stop Detection, 47–8
Stereo Pairs, 131–2
Storage:
  location, 13–14, 23–4, 28–30
  space, 6, 33–4
Subclips, 36–7, 87–8
Superimpose, 66
Sync, 80–1

Tape backups, 140
Televisions:
  aspect ratio, 9, 28
  display area, 113
Text, 101, 123–7, 149
Three Point Editing, 64–5
Three way dissolves, 121–2

Thumbnails, 45
Timecodes, 33, 37–8
Timeline, 16–17, 60
  expansion/contraction, 62, 79–80, 135
  red bar, 85
Titles, 123–7
Titlesafe, 113, 127
Tools, 16–17, 69–71
  Arrow, 70, 75–6
  Cinema, 148–9
  Magnifier, 70, 79–80
  Pen, 70
  Pointer, 70, 73
  Razorblade, 70, 78–9
Tracks:
  active, 67
  adding, 69, 100, 108, 136–8
  CDs, 39, 137
  control track, 37–8
  deleting, 69
  editing, 67–9, 73, 79, 93
  linked selection, 71–3
  locked, 68–9, 73, 79, 93
  multiple, 108
  selecting, 70
  sound, 149
Transitions:
  audio, 135–6
  video, 101, 102, 103–4

Undo, 71

Variable bit rate (VBR) encoding, 149
Video:
  compositing, 107–27
  copying effects, 122–3
  effects, 99–127
  filters, 101, 102, 104–6
  formats, 8–9, 13, 27, 30–1
  generators, 101, 102, 123–7
  layered effects, 100–2
  moving image positions, 117–20
  multi-layered effects, 114–22
  real-time effects, 86–7
  rendering, 84–6
  storage requirements, 6
  titles, 123–7

Video (*continued*)
   tracks, 67–9
   transitions, 101, 102, 103–4
Viewer, 16–17
Viewing clips, 44–5
VTR Controller, 33

Widescreen anamorphic
   format, 9
Windows:
   Capture, 32–6
   layout, 16–20
Wipes, 101